D1561484

Yale
Agrarian
Studies
Series

James C. Scott, *series editor*

"The Agrarian Studies Series at Yale University Press seeks to publish outstanding and original interdisciplinary work on agriculture and rural society—for any period, in any location. Works of daring that question existing paradigms and fill abstract categories with the lived-experience of rural people are especially encouraged."

James C. Scott, *Series Editor*

Christiana Payne, *Toil and Plenty: Images of the Agricultural Landscape in England, 1780–1890* (1993)

Brian Donahue, *Reclaiming the Commons: Community Farms and Forests in a New England Town* (1999)

James Scott, *Seeing Like a State: How Certain Schemes to Improve the Human Condition Have Failed* (1999)

Tamara L. Whited, *Forests and Peasant Politics in Modern France* (2000)

Nina Bhatt and James C. Scott, *Agrarian Studies: Synthetic Work at the Cutting Edge* (2001)

Peter Boomgaard, *Frontiers of Fear: Tigers and People in the Malay World, 1600–1950* (2001)

Janet Vorwald Dohner, *The Encyclopedia of Historic and Endangered Livestock and Poultry Breeds* (2002)

Deborah Fitzgerald, *Every Farm a Factory: The Industrial Ideal in American Agriculture* (2003)

Stephen B. Brush, *Farmer's Bounty: Locating Crop Diversity in the Contemporary World* (2004)

Brian Donahue, *The Great Meadow: Farmers and the Land in Colonial Concord* (2004)

J. Gary Taylor and Patricia J. Scharlin, *Smart Alliance: How a Global Corporation and Environmental Activists Transformed a Tarnished Brand* (2004)

Raymond L. Bryant, *Nongovernmental Organizations in Environmental Struggles: Politics and the Making of Moral Capital in the Philippines* 2005)

Michael Goldman, *Imperial Nature: The World Bank and Struggles for Social Justice in the Age of Globalization* (2005)

Edward Friedman, Paul G. Pickowicz, and Mark Selden, *Revolution, Resistance, and Reform in Village China* (2005)

Nongovernmental Organizations in Environmental Struggles

Politics and the Making of Moral Capital in the Philippines

Raymond L. Bryant

Yale University Press New Haven & London

Set in Minion type by Keystone Typesetting, Inc.
Printed in the United States of America.

Library of Congress Cataloging-in-Publication Data

Bryant, Raymond L., 1961–
Nongovernmental organizations in environmental struggles : politics and the making of moral capital in the Philippines / Raymond L. Bryant.
p. cm. — (Yale agrarian studies series)
Includes bibliographical references and index.
ISBN 0-300-10659-9 (cloth : alk. paper)
1. Political ecology—Philippines. 2. Non-governmental organizations—Philippines.
3. Business ethics—Philippines. I. Title. II. Yale agrarian studies.
JA75.8B788 2005
304.2′09599—dc22

2004059853

A catalogue record for this book is available from the British Library.

The paper in this book meets the guidelines for permanence and durability of the Committee on Production Guidelines for Book Longevity of the Council on Library Resources.

10 9 8 7 6 5 4 3 2 1

For Shanti

Contents

Acknowledgments

A book that can be traced back many years is bound to leave a long trail of intellectual and other debts. As my debts span numerous countries, a multitude of institutions, and countless individuals, it is not feasible to name them all here. The fieldwork itself was made possible by the financial assistance of the Nuffield Foundation and the Economic and Social Research Council in the United Kingdom.

If this is a book that seeks to promote theoretical understanding around notions of moral capital and moral entrepreneurship in relation to nongovernmental organizations (NGOs), it is nonetheless a work that is embedded in empirical research conducted in 1996 and 1997 in the Philippines. I am thus very grateful to many individuals and institutions that helped facilitate that research. My main research base was at the Environmental Research Division (subsequently renamed Environmental Science for Social Change) located at the Manila Observatory on the edge of the Ateneo de Manila University campus. Peter Walpole has long been the guiding force at the ERD/ESSC and I owe him a particular debt of gratitude. He not only ensured that my accommodation and office needs were amply met, but also provided me with critical introductions into the Philippine NGO world without which this study would have been impossible. Peter also managed to fit in the time to offer constructive criticism of both the theoretical and empirical foundations of this book. He was not alone. Other staff at the ERD provided intellectual and/or practical support. I am especially indebted to Gilbert Braganza, who helped get me started in Manila and who provided thoughtful and always constructive commentary on both the moral capital perspective and the Philippine case studies. Mrs. Cornelio meanwhile provided all of the administrative support that I could have wished for as I sought to deal with the inevitably complicated logistics of fieldwork. Thanks also go too to the British Council for its willingness to finance the development of an institutional link between King's College Geography and ERD. This link enabled me to conduct initial research in Manila in January 1996 and has been a fruitful conduit for research collaboration between the two institutions more generally.

As I began to make connections in the Philippine NGO sector, I was influenced by what a number of long-standing observers and participants have had to say. Many of these individuals patiently answered my questions and suggested further areas and issues for research. Some of them debated the premises of this study. Others were sufficiently intrigued (or merely polite) to accede to a second interview. All of those who permitted a formal interview are listed at the end of this volume (see Interviews). I wish here to thank all of them for their time, patience, understanding, and enthusiasm! Special thanks for wise advice go to Patrick Dugan, Marites Dañguilan Vitug, Mary Racelis, Alan Alegre, Karina Constantino-David, Fernando Aldaba, and Delfin Ganapin.

Part of the research involved short, focused visits to selected locations outside of Manila. While my study is focused on two Manila-based Philippine NGOs, the nature of their work naturally led me to wish to visit communities where they have worked. I am grateful to various local communities and their representatives who permitted me to visit them in and around Bolinao, Mount Isarog, Didipio, and Coron Island so as to be able to gain a better appreciation of the work of the two NGOs in "the field." Their hospitality helped me understand local conditions and NGO work there. Thanks too go to Aida Granert at the Soil and Water Conservation Foundation for helping to coordinate my visit to Cebu City.

One of my biggest debts of gratitude is to the two Philippine NGOs that form the empirical core of this study. I was delighted that the Haribon Foundation and the Philippine Association for Intercultural Development (PAFID) agreed to be a part of my research project. These two pioneering NGOs are renowned for their work on environmental and indigenous rights issues respectively—and staff members are all subject to major pressures on their time. Yet, not only did both organizations agree to subject themselves to what must have seemed like a never-ending series of questions, they did so with much patience and good humor to boot. Both in Manila and in the various locations where they work, the staff of these organizations provided me with invaluable help. I wish to thank all of those who additionally agreed to formal interviews and in particular to Ed Tongson and Cristi Nozawa at Haribon and to Delbert Rice, Dave De Vera, Andre Romero, and Lourdes Amos at the PAFID for their general assistance.

There were two other essential ingredients to the successful completion of my Philippine fieldwork. One was the expert research assistance of Joy Abelardo and Tonette Lapus. Not only did they provide logistical coordination surrounding interviews and fieldtrips (including occasional translation), but they also helped uncover a wealth of Philippine secondary and newspaper material that has greatly enriched this study. Their expertise was pivotal to the progress of the research and I will always be grateful for their support and enthusiasm. The other ingredient was my good fortune to stay with the Kornerup family in Quezon City. As the most welcoming of hosts, they provided me with invaluable pointers on Philippine society and hours of interesting discussion.

In the United Kingdom, I have benefited from the intellectual support and insight of a variety of people. Preliminary results were presented at the Association of Southeast Asian Studies UK Conference in March 1997 at the University of Hull. The arguments were subsequently elaborated at the Center of Southeast Asian Studies of the School of Oriental and African Studies (SOAS) in November 1998 and at the Department of Geography, University of Cambridge, in November 1999. My thanks to all of the participants who provided useful feedback as well as to Tim Forsyth, Philip Stott, Gerard Clarke, Robert Taylor, David Potter, Philippe Le Billon, James Putzel, James Manor, and John Sidel for their advice and/or insights over the years. At my own institution, I have benefited from discussions with a number of my postgraduate students, including notably Richard Gauld, Sinead Bailey, "Otto" Jeng-di Lee, and Karen Lawrence. The latter in particular has been especially helpful in providing constructive criticism and invaluable Philippine materials over the years. All of them have displayed an endearing *un*willingness to accept at face value what I and other King's academics say. My colleagues at King's have also challenged me in various ways over the years to rethink my assumptions about NGOs and political ecology —albeit not always to their liking! Special thanks go to Keith Hoggart, Michael Redclift, John Thornes, Linda Newson, Chris Hamnett, Geoff Wilson, Margaret Byron, and James Defilippis. Shatish Kundaiker has been very patient with my (naïve) computing questions. Roma Beaumont and Carolyn Megan have been much too kind in helping with the figures in this book. I have long marveled at their ability to translate my squiggles into something intelligible to a wider audience. Alison Greene has provided me with invaluable administrative support over the years. Without her help this book would certainly have been even longer in the making than it was.

Various colleagues in the United States have been patient with me as I have sought to relate my theoretical concerns to empirical conditions in the Philippines. Various portions of the argument have been tested at annual meetings of the Association of American Geographers held in Boston (in 1998), Pittsburgh (in 2000), and New York City (in 2001). For feedback at these meetings or simply useful exchanges, I would like to thank Richard Peet, Anthony Bebbington, Michael Goodman, Karl Zimmerer, Lucy Jarosz, Cindi Katz, Nayna Jhaveri, Arturo Escobar, and Joshua Muldavin. Some of the ideas contained herein were also tested through papers presented at various university seminars. At the Department of Geosciences at the University of Massachusetts, Amherst, I wish to thank Stan Stephens and James Boyce for their helpful remarks, and I am especially indebted to James Hafner for his constructive input and support. I thank staff and students at the School of Geography, Clark University, for their challenging response to my presentation, in particular Dianne Rocheleau and Thomas Ponniah, who kindly facilitated my visit and also made it a memorable one. My thinking on moral geographies and conservation has also been shaped by the wonderful feedback and searching criticism of the students who attended my lecture at the School of Forestry and Environmental Studies, Yale University. I am grateful to Michael Dove for the kind invitation to give this lecture and for being a generous and thought-provoking host. Finally, my thanks go to all of those who are involved with the Yale Program in Agrarian Studies (and in particular James Scott) for inviting me to their colloquium series. I would like to thank all of the participants for their very helpful comments on my paper and especially Donald Moore for detailed feedback.

James Scott has been pivotal in the making of this book in other ways. For many years he has been a source of inspiration to me in terms of thinking about some of the complexities of power relations between the weak and the strong—and all of those who fall somewhere in between. He was also the guiding force in leading me to think about Yale University Press as a logical "home" for my work on NGOs, moral capital, and the Philippines. On both counts I thank him very much. At Yale University Press, I would like to thank my editor Jean Thomson Black, who has shown much patience in dealing with my many questions as well as the innumerable delays encountered along the way. My thanks go as well to two anonymous referees whose detailed and constructive comments on the initial manuscript have resulted in a stronger final version. I am also grateful to Marie Blanchard for her thorough edit of the manuscript.

This book took much longer than planned. The research and writing inched forward at what seemed to me at least to be a painfully slow rate. The routines of teaching and administration played a part here, but the rigors of domestic life also intervened. I like to think, though, that this is a better book for all of that enforced delay. Above all, I have Shanti to thank for providing critical intellectual and emotional support at every twist and turn in this inquiry. Her comments on successive drafts were invaluable even as she reminded me that there was more to life than academic inquiry. Indeed, our children, Priya and Kiran, born during the work on *Nongovernmental Organizations in Environmental Struggles,* have continually demonstrated to me that there are more important things than writing a book.

Introduction

This book reflects a longstanding personal interest in nongovernmental organizations (NGOs) and environmental struggles. That interest initially found expression through support for environmental organizations in the First World. In the 1990s I related this experience to research on Third World politicized environments, with an initial attempt appearing in the book *Third World Political Ecology* (Bryant and Bailey 1997). The conclusion there was that the social and political impact of environmental NGOs was decidedly ambiguous. There was certainly evidence to suggest that they had prompted powerful actors such as states, businesses, and international financial institutions to change their practice. Yet there was also evidence of tension among NGOs over funding, philosophy, and strategy, notably vis-à-vis local communities. Such tension was also linked to the considerable diversity of entities labeled "nongovernmental organizations"—a diversity that partly explains the fierce debate that surrounds the NGO sector. For the purposes of this book, though, I adopt the definition provided by Gerard Clarke (1998: 2–3, italics in original): NGOs are "*private, non-profit, professional organizations with a distinctive legal character, concerned with public welfare goals.*"

Barely had ink dried on paper, however, before I became dissatisfied with my initial assessment of NGO impacts. What was lacking there—and in many contemporary accounts—was any sense of *why* such ambiguity occurred, let alone *how* organizations might accommodate themselves to the vicissitudes of NGO life. How do NGOs persuade others to change their views and practices? Indeed, how do they do so under habitual conditions of financial insecurity? Might these situations even be connected? It struck me that an exploration of how NGOs seek to empower themselves in order to pursue their social aims and objectives was urgently needed. The opportunity to conduct fieldwork on strategizing by Philippine NGOs in environmental struggles seemed a perfect opportunity.

Two incidents close to home—one widely reported and one quite "trivial"—provided further grist for the mill. The first was the Brent Spar incident in northern Europe. Here, the NGO Greenpeace fought the plan of Shell Oil to dispose of decommissioned North Sea oil platforms and storage buoys, of which Brent Spar was the first. The company wanted to sink them in the ocean, while the NGO preferred "less ecologically disruptive" onshore disposal. The battle intensified in 1995. The British government threw its support behind Shell. Yet Greenpeace was able to persuade the media to cover the issue. Through this medium, the organization appealed to consumers to boycott Shell products. This tactic was successful. There was a fall in sales at the pump and even an arson attack on premises in Germany. Panic set in at headquarters. The company decided to abandon ocean disposal even though most scientific evidence presented it as the best option under the known circumstances. Greenpeace thereby triumphed in one of its more successful political "soap operas" (Dickson and McCulloch 1996; Smith 2000). What interested me about this struggle was the way in which moral perceptions influenced a political outcome. It struck me that the advantage of the NGO here was neither superior political resources (money, political access) nor scientific knowledge. Rather it was superior moral credibility with many reporters and consumers that enabled it to persuade people to change behavior in a way likely to damage an opponent grounded in the "moral austerity" of conventional environmental decision-making (Gillroy and Bowersox 2002).

The second incident was equally revealing. It involved a Greenpeace conference held in London in October 1997 designed to build an alliance with business on the environment. The most interesting thing here was the context—a point seized upon by a reporter in attendance. We are told that the meeting occurred in a luxury hotel and that Greenpeace leaders were dressed, as were their business counterparts, in suits. Indeed, "at times, it was hard to tell who was who" (Lean 1997: 8). Ostensibly trivial details—suit wearing, luxury hotel—were noted to generate an unflattering portrait of Greenpeace leaders hobnobbing with business executives. This incident—alongside other reports on NGO "appearance"—pointed to an abiding tension associated with this sort of organization (Karacs 1999). The newspaper report appeared to summarize a key ambiguity faced by many NGOs. By speaking to elites thus, organizations like Greenpeace run the risk of being seen to be "too close for comfort" to those with whom they are popularly identified as being in conflict (Hulme and Edwards 1997). An NGO might be morally discredited simply because it acted "out of character." Moral credibility and prestige thus appeared *simultaneously* to be a resource leading to possible NGO empowerment and a set of double-edged expectations of "correct" conduct.

These ruminations accompanied me on an initial field trip to the Philippines in early 1996. With colleagues at the Environmental Research Division (ERD) of the Manila Observatory, I elaborated my research project. Yet, as my head buzzed with names and logistical details, I kept returning to this question of moral perceptions, strategic behavior, and NGO empowerment. As I learned about

Philippine NGOs, I discussed my plans with NGO leaders, academics, state officials, and donor representatives. The result was a clarification of short- and long-term issues surrounding the development of the NGO sector in general and the environmental NGO sector in particular. The discussions also convinced me that the Philippines would be an ideal country in which to pursue my emerging concerns, in that it had one of the largest and most sophisticated NGO sectors in the world. Part of that sophistication, it struck me, was associated with the ways in which organizations constructed moral agendas as well as how those agendas were received and indeed influenced by diverse audiences.

Funding worries were certainly involved in this saga. Yet there was more at stake in understanding Philippine NGOs than simply the realization that funding allocations influenced what they did. While not denying the money "merry-go-round," I was nonetheless keen to further assess NGO strategic behavior and empowerment. To what extent were such behavior and capacity linked to moral visions and missions espoused by NGO staff members? What was the practical political significance of these manifestos?

An epiphany occurred as I traveled between interviews. It struck me that NGO strategic behavior was notably attuned to the need to boost moral standing with such actors as local communities, donors, the media, or state agencies. This process I labeled a quest for *moral capital*. The term was designed to capture my sense that NGOs should be seen as involved in a complicated social and political process of "resource mobilization"—albeit one in which cultural dynamics and strategic rationality are inextricably intertwined.

This connection was not widely acknowledged in the literature. There was much comment on the relative moral stature of NGOs and social movements—a saint-versus-sinner type of debate. Yet this normative enterprise was different from my concerns (see also Johnston and Klandermans 1995; Lahusen 1996; McAdam et al. 1996;

Jasper 1997). I was not that interested in whether NGOs acquitted themselves in a "morally propitious" manner. Rather, I wanted to gauge the significance of the idea that the strategic behavior and empowerment of NGOs was associated with the moral perceptions that actors had about them. Mine was thus conceived of as an exercise in description, albeit with potentially wide-reaching implications for how NGO action is understood.

As I related NGO action to a quest for moral capital, it became necessary to explain selected practices in a new way. That an NGO devises a multifaceted strategic approach is not new. That it may do so with an eye to boosting its moral reputation is a different matter. The possible novelty here was to be found in a reassessment of diverse NGO activities. Such things as the calibrated criticism of state agencies, the push for private citizen (as opposed to donor) funding, the individual benefits of coalition work, the defense of area-based reputation, and turf wars all seemed of a piece. They appeared to be outcomes of "entrepreneurial" action by NGOs keen to acquire moral capital with diverse partners.

Otherwise "irrational" action became understandable too. It now made sense that an NGO might reject funding—even if this step alienated powerful actors—when its moral standing with especially valued partners was at stake. It now made sense that an organization might remain in an area despite adverse financial and political events (including physical violence against employees). It made sense too that an NGO would persist in criticizing state policy or practice even if that meant that it suffered intense political heat in return.

Yet I did not wish thereby to romanticize NGOs. I realized, as research got under way, that the quest for moral capital was fraught with political *and* ethical ambiguity. That an organization pursues moral capital in an entrepreneurial fashion is no guarantee that awkward tradeoffs will be avoided. To the contrary, the need to balance a concern for moral standing with the need to attend to the daily exigencies of life often places organizations in a bind. Doing "good" may be a reward in itself. However, that reward is rarely sufficient to enable the effective promotion of a vision and mission. There is thus a need to promote an image of "doing good" even as actions such as the acceptance of business funding may imperil that image. Moral capital may be a potentially important resource. Yet there is a terrible irony here: moral capital is as much about vulnerability as it is about opportunity. As Kane (2001: 3) observes, vulnerability is "a consequence of the fact that moral capital exists only through people's moral judgments and appraisals and is thus dependent on the perceptions available to them." In short, there is no guarantee that strategic behavior will result in high moral standing. Indeed, being seen to be too blatant in the effort to "curry favor" might even prove counterproductive for an NGO. There is a risk therefore of being seen to be *too* calculating in the pursuit of moral capital such that a net *loss* of capital occurs (Hjelmar 1996; Fisher 1997).

Why bother, then, with a quest for moral capital, given these uncertainties? The rationale is not for the moral capital itself. This resource is a means to an end rather than an end in itself in a context where ends themselves may shift over time. It is an intermediate "good" designed to unlock diverse political, economic, and social benefits. It is a resource that is helpful only when pressed into "useful service" (Kane 2001: 7). This process of putting moral capital into useful service is akin to NGO empowerment. What goals are involved here? The NGO with moral capital may be in a good position to acquire funding, earn media respect, win public support, enjoy policy influence, and receive local community backing. *None* of these benefits come automatically, but they *are* often attainable with moral capital. In short, there is an incentive to think like a moral entrepreneur.

The importance of NGO strategic action thus needs to be kept in full view. Such action requires in turn assessment of three interrelated strategies. The first is that of *political strategy,* or the manner in which an NGO addresses its political relationships. This strategy, covered in chapter 4, notably comprises critical engagement with state agencies and constructive engagement with local communities. Such multifaceted engagement is frequently noted in the NGO literature (Alegre 1996; Fisher 1998; Mercer 2002), yet a moral capital perspective facilitates new insights into how and why such behavior occurs.

The second component is that of *financial strategy*. Here, I assess the way in which an organization addresses financial need while accounting for noneconomic considerations. As chapter 5 suggests, a strong impetus to promote financial autonomy can determine where an NGO seeks funding and what it does once it has found it. Indeed, the quest for moral capital may lead to "sub-optimal" economic outcomes, especially in the short term. An organization may "waste" energy on "marginal" fund-raising even as it may feel impelled occasionally to reject money already given.

Such political and financial strategies are linked to a third component—*territorial strategy*. Yet chapter 6 suggests that the latter is not simply the spatial expression of the other two strategies. The effort to map the mission can take on a life of its own such that spatial and territorial dynamics add an edge to the quest for moral capital. This is seen in the pursuit of area-based reputations as a means of attaining spatial economies of scale in the dissemination of a good name. It can also be seen in the lengths to which an NGO may go to defend "soft territoriality"—zones of influence in which moral capital is acquired and used. That other NGOs are often rival claimants underscores that NGO solidarity cannot be taken for granted, especially in a world where NGOs think like moral entrepreneurs.

The empirical portion of the book is therefore an elaboration and assessment of arguments that relate political, financial, and territorial strategizing by NGOs to a quest for moral capital. The analysis is preceded by a short overview of Philippine NGOs as well as the two case study NGOs (chapter 3). This book thus ought to appeal both to readers interested in the Philippines and a wider group that seeks new understandings of NGO strategic behavior and empowerment. Indeed, the novel theoretical framework is designed to provoke debate and comparative inquiry integral to theory building. To this end, chapters 1 and 2 explain the assumptions and propositions that comprise the moral capital perspective as developed in this book. Chapter 1 explores the chain of assumptions that underpin this perspective, relating an understanding of morality and altruism to the notions of reputation and power. In chapter 2, theorizing on noneconomic forms of capital by Bourdieu and others is considered as a precursor to an elaboration of the moral capital perspective. A short discussion of the methodology used closes that chapter.

The evidence thus acquired is based on the strategic behavior of two reform-minded NGOs operating under contingent democratic conditions. They are both "mature" medium-size organizations that are pioneers in their fields. How applicable, then, is the moral capital perspective, given the diversity of NGOs in the world? If this question is an invitation for future comparative research, I do consider it briefly in chapter 7, where a research agenda is sketched. However tentative, that discussion suggests that moral capital *might* be a resource whose utility is appreciated and impact felt beyond the confines of the NGO worlds described herein.

CHAPTER 1

Capitalizing on the Midas Touch

To the extent that NGOs are culturally resourceful, they can turn moral concerns and perceptions into a capacity to act. That they may be seen to behave altruistically arguably enhances an ability to promote a favored mission. Yet to describe NGOs thus does not help us to understand *how* they translate a perceived high-mindedness into action. Indeed, there may even be a tension between thinking of NGOs as "benevolent visionaries" and thinking about them as "hard-nosed pragmatists" that must compromise.

This tension resonates in the literature. A "utopian" school is epitomized by the work of Korten (1990). His argument is that the main contribution of an NGO is to facilitate "people-centered development" by coordinating locally responsive networks of actors. These organizations promote a "transformation agenda" of justice, inclusiveness, and sustainability. If their strategy has shifted over time, this is mainly in response to changing local needs rather than the demands of elites. The energy propelling NGOs to the forefront of "associational revolution" has thus been about "making a difference" in the lives of the downtrodden (Ekins 1992; Edwards and Hulme 1992; Salamon 1994; Fowler 1997).

A "dystopian" school dwells on the dark side of NGO experience.

This perspective, seen in work by Smillie (1995) and Sogge (1996a), would puncture "myths" surrounding NGOs. They probe the compromises that organizations make with powerful elites in order to function—compromises that may even undermine NGO ideals. This bleak picture is rounded out with an assessment of how altruism and solidarity is subverted due to organizational self-interest and competition. Indeed, some argue that when NGOs get "too close for comfort" to powerful political and economic elites, "like Icarus before them, [they] may plummet to the ground when the heat of the donors melts the wax in their wings" (Hulme and Edwards 1997b: 284; see also Najam 1996; Tvedt 1998).

This summary clearly oversimplifies a complex picture. Indeed, it bears reiteration that NGOs are a diverse lot. For example, they are differentiated by size, function, and scope of operation, let alone philosophy, ideology, or national origin (Farrington et al. 1993; Meyer 1999; Mercer 2002). Such diversity has a bearing on the possible wider applicability of the moral capital perspective—a question touched on in the introduction and one that is considered in more detail in chapter 7. Here, though, this summary helps introduce the view of NGOs as a broad "class" of actors characterized by an intermediate social position that entails much ambiguity. This is both their strength and their weakness. Yet, in general terms, many of these organizations seem "Janus-like" in that power seems to be derived from actions that sometimes appear visionary (as the utopian school believes) and other times seem "pragmatic" (as the dystopian school argues). There is rarely a hard and fast distinction here. Still, much NGO cultural resourcefulness resides in working with or against people to shift this process to their liking. Such strategic thinking and behavior is the focus of this book.

Ambiguity over behavior crops up in related debates in the social movement literature. According to resource mobilization theory (RMT), articulated notably by McCarthy and Zald (1973; see also

Zald and McCarthy 1987), strategic thinking by social movement organizations (SMOs) counts most, since much of what they do is to mobilize political and economic resources to resolve social grievances. RMT has shown nuance in its argument in recent years as it emphasizes how mobilizing strategies are linked to cultural "framing." Still, its explanatory power remains linked to assessments of the practical dilemmas and choices that confront SMOs on a daily basis, even if it has been attacked thereby for being narrowly rationalist in outlook (Ferree 1992; McAdam et al. 1996; Della Porta and Diani 1999).

In contrast, new social movement (NSM) theorists like Touraine (1981) and Melucci (1989; 1996) examine cultural dynamics, focusing attention on how and why collective identity is articulated. Notice is certainly given to day-to-day realities, but the claim that movements challenge "cultural codes" through words and deeds bespeaks a literature inclined to radical interpretation of these social formations. NSM theorizing is thus concerned with the potential for transformation associated with these moral and cultural visionaries (DeLuca 1999; Schlosberg 1999). These debates draw attention to the manner in which SMO and NGO power may be linked simultaneously to visionary practices (power through moral visioning) and "pragmatic" actions (power through social maneuvering). The two aspects are certainly connected. Part prophet and part hustler, many organizations seem to derive power from an ability to "speak the truth" to different actors in different social settings. They use cultural skills to acquire the resources to act even as sometimes, through the shrewd use of those resources, they may alter cultural norms and practices (Foweraker 1995; Alvarez et al. 1998; Smillie and Hailey 2001).

Environmental struggles test that resourcefulness. As the political-ecology literature demonstrates, many NGOs rail against a world dominated by global capital and associated political practices. The ramifications here include widening income inequality and pervasive environmental disruption. The creation of winners and losers appears to be an integral part of a prevailing political economy based on social and environmental contradiction (Ecologist 1993; Harvey 1996; Bryant and Bailey 1997; Braun and Castree 1998; Zimmerer and Bassett 2003; Peet and Watts 2004). That such mayhem furnishes many NGOs with a raison d'être is inadequately emphasized in this literature. The uneven development of global capitalism leaves an ever-shifting sociospatial and biophysical imprint that has enabled the rapid rise to prominence of NGOs and social movements around the world. It is not that NGOs are inevitably rooted in an anticapitalist perspective. Rather, it is that the social and environmental issues to which they respond are notably a manifestation of capitalist and allied state practices (Yearley 1996; O'Connor 1998; Keck and Sikkink 1998; Brosius 1999; Mercer 2003; Routledge 2003).

Environmental NGOs race to protect biodiversity "hotspots" before they are destroyed by unchecked resource extraction. This effort is often coupled with the promotion of "green" capitalist practices such as ecotourism or biodiversity "prospecting." Meanwhile, development NGOs help those who become socially marginal as a result of unequal development aided and abetted by states (McAfee 1999; Bryant 2002a; Mercer 2002; Sundberg 2003). As I document with reference to the Philippines, environmental struggles mean that NGOs are pulled in different directions as moral visions sometimes clash with more immediate dilemmas of social and political interaction (Allahyari 2000). Yet many organizations not only survive but actually prosper. That such power may be acquired even under difficult circumstances points to a strategic logic and behavior that is my main concern. To make sense of such strategizing is first to consider a chain of assumptions that links contingent NGO identities and purposes to perceptions of moral and altruistic action, and which then relates those perceptions to the question of NGO reputations and empowerment.

NGOs as Moral and Altruistic Actors

Because moral issues address crucial questions of human identity, it is not surprising that they exercise the minds and passions of many people at one time or another in their lives. Indeed, as Smith (2000: 1) argues, moral reasoning is "part of our distinctively human nature." Yet, to recognize the ubiquity of moral reflection does not detract from a key feature of many NGOs—namely, that they are created specifically to advance a moral vision. Many of them pursue that vision primarily on behalf of "others": disadvantaged people or endangered flora and fauna, for example. Here, then, is a possible basis for seeing them as altruistic. To suggest that NGOs may be seen as moral and altruistic actors, however, raises questions about the meaning of moral and altruistic conduct.

To claim that NGOs promote the "good society" does not mean they agree on its features. Their concerns are diverse and range from questions of global peace and ecology to issues of personal identity and social regulation (Anheier and Salamon 1998; Goodwin et al. 2001; Polletta 2002; Bryant and Jarosz 2004). Those concerns are also linked to an array of political ideologies and practices including conventional left-right divisions as well as deep/light green splits (A. Scott 1990; Dalton 1994; Heyzer et al. 1995; Doyle and McEachern 2001; Hart 2001; Connelly and Smith 2003; Routledge 2003). Such diversity is not surprising. To dwell too much on it, though, is to miss the point that these organizations are defined as a group in part because they often privilege moral concerns. There seems to be a moral raison d'être that tends to set them apart from other actors such as business corporations and states. Further, as many NGOs are seen to pursue moral concerns, they can acquire a good name vital to their ability to effect change.

Still, two conditions must hold if moral concerns are to be a basis for empowerment. These are (1) that a sense of morality is indispensable to society, and (2) that morality needs to be understood as a social process. Without the former, the moral basis of NGO concerns is socially irrelevant, thereby reducing this actor's ability to acquire power. The absence of a view of morality as a social process calls into question the need for an organized social response to moral issues, once again curtailing prospects of NGO empowerment. It is worthwhile, then, to briefly consider these two conditions.

Strong grounds exist for thinking that societal cohesion is possible only when basic moral principles apply. Indeed, scholars from diverse disciplinary backgrounds have long made this argument in linking moral issues to questions about human identity and collective practice (e.g., Wolfe 1989; Parry and Bloch 1989; Benhabib 1992; Goodin 1992; Howell 1997; Smith 1998, 2000). There are nonetheless sharp differences over the precise role that morality ought to play in the regulation of human behavior (increasingly seen by some as linked to interaction with the biophysical environment; see Dobson 2000). A standard division here is between what is termed the domains of the Right and of the Good. Thus, one view treats morality in a narrow sense in that it links morality with the domain of the Right, understood as comprising principles of justice pertaining to rights and obligations. Here, the aim is to constrain human conduct so that basic human interests are safeguarded (Lukes 1991). The narrow view is often based on the assumption that humans are in perpetual competition and conflict due to limited resources and sympathies. In contrast, the broad sense of morality encompasses not only the domain of the Right, but also the domain of the Good. In this case, there is an "all-inclusive theory of conduct" at stake (John Mackie, cited in Lukes 1991: 21). In this book, it is not the relative merit of these views that is important—let alone the aspiration to rational self-sufficiency in the face of social contingency that both views reflect (Nussbaum 1986). Rather, it is that there is a consensus that *some* morality is essential to the operation of human society. While these (and other) views may judge a society to be more or less morally situated, the crucial point for our purposes here is that such evaluation is widely seen to be essential to social well-being even if it may raise thorny questions about the commensurability of values in the process (Radin 1996; Gudeman 2001).

However fraught and ambiguous the process might be, much of what NGOs do is nonetheless precisely concerned with moral evaluation and assertion. They seek to publicize and rectify perceived discrepancies between what they believe is Right and Good in theory and what occurs in practice. Indeed, these organizations probe how societies (or parts thereof) may not live up to even basic moral principles. Many NGOs thus promote an agenda in which moral perspectives loom large even as that effort poses an inescapable risk of slipping into unsavory and self-defeating moralism—what has been dubbed "moraline drift" (Bennett 2002; see also Fisher 1997; Brosius 1999).

While a minimum level of morality may be vital to human welfare, it does not necessarily follow that NGOs or other types of organization are the best means to attain that goal. It could be argued, for example, that it is for the individual to regulate his or her behavior in keeping with moral principles. Here, the attainment of moral goals is ensured via individual self-interest and initiative. This exaltation of the individual is, of course, at the heart of an influential if extreme variant of rational choice theory linked to the "Chicago school" of economists and policy practitioners (e.g., Becker 1981). Such thinking, in turn, has been widely attacked in the literature (e.g., Radin 1996; Zelizer 1997; Gudeman 2001).

My concern here is with the view of morality as the exclusive preserve of the utility-maximizing individual. At first glance, to specify moral action thus would be simply to acknowledge the seemingly incontrovertible fact that it is the human body that is "the irreducible basis for understanding" (Harvey 1998: 40). On closer inspection, though, there is a basic tension in this argument. Maximizing individual utility through the market is seen to guarantee a moral society because, as individuals pursue their interests, they simultaneously build collective prosperity. Yet, market "efficiency" presupposes individuals guided by moral principles inculcated outside of market relations (through religious instruction, for instance). For individuals to be otherwise inclined is to call into question the ability of buyers and sellers to trust one another. Since trust is essential to market operations, its absence would jeopardize effectiveness. Yet there are no guarantees that moral education will happen and certainly no sense of how moral behavior in the market itself is to be affirmed in the face of competition. Nor is there an indication of how, even in a society of moral individuals, individual moral action will add up to something that meets even minimal collective obligations. The latter may never be achievable through individual action alone. By reducing morality to individual self-interest, this form of rational choice thinking falls into a Hobbesian trap in which there is neither an adequate social mechanism nor a public-minded spirit by which the "war of all against all" can be overcome (Jordan 1989; Wolfe 1989; Plant 1992).

There is thus a need to view morality as a social process. Thinking about morality in this way views interests as socially constituted. In the process, an opportunity arises for individuals or groups (such as NGOs) to "play the role of moral entrepreneur: to draw attention to new needs and to encourage others to act together in some new combination" (Jordan 1989: 170). The need for moral entrepreneurs seems especially great when action is required on behalf of socially marginal groups or even endangered nonhuman species (Allahyari 2000; Chaloupka 2002). Yet, if many NGOs derive purpose from highlighting a perceived gap in theory and practice concerning the Right and the Good, it is only because they work in societies in which morality is defined socially that they have purchase on moral debates. "Moral limits to the market" may present NGOs with the opportunity to be—and to be seen to be—moral actors (Plant 1992).

The role of NGOs as moral concerns is related in complex ways to the question of class. Here it needs to be recognized that morality itself is not a disembodied thinking process but is anchored in material struggle. This point can be illustrated when considering the phrases "middle-class radicalism" and "petit-bourgeois moralizing." It is commonly observed that most NGOs derive support from the middle class (Eder 1993; Gregorio-Medel 1993; Yearley 1996). Their actions tend to be inflected with the anxieties and interests of that class too. As Eder (1993: 156) notes, such radicalism is primarily moral since the petit bourgeoisie is "predestined by its social position to mobilize moralists." He adds that the distinctive position of the petit bourgeoisie—neither bourgeoisie nor proletariat—is an intermediary position that simultaneously renders it a keen supporter and beneficiary of capitalism, but ever a possible victim of that self-transforming economic system. A petit bourgeois habitus (or collective disposition) centers on the assertion of moral positions of Right and Good in order to maintain class distinction (Bourdieu 1984). The favored petit-bourgeois action is the moral crusade in which "the difference between moral ideal and social reality becomes the motivating force of collective protest" (Eder 1993: 149).

To reflect on the middle-class origins of NGO moral action is to clarify aspects of wider organizational identity. It is also to shed light on why these entities can be so preoccupied with and publicly linked to moral issues (Wapner 1996). Yet it is wrong to conclude that the moral concerns of many NGOs are confined *only* to the petit bourgeoisie. That class may play a special role in articulating them, but moral concerns enjoy a wider social circulation (Allahyari 2000; Goodwin et al. 2001). In the process, they acquire a currency extending well beyond their origins, even sometimes crisscrossing political and cultural boundaries (Bauman 1993; Low and Gleeson 1998; Routledge 2003).

That moral concerns circulate and acquire meaning in different social settings raises the question of the social construction of moral arguments. Moral concerns acquire their meaning from the way in which disparate ideas and values are brought together in specific discourses. To talk about moral discourses is to enter a world in which actors pitch their messages using particular rhetorical idioms replete with evocative motifs and expressed through distinctive claims-making styles (Ibarra and Kitsuse 1993; Hannigan 1995; Lara 1998; Kurtz 2003; Martin 2003). To take but one example, environmental NGOs tend to adopt a rhetoric of loss associated with the idea that a "pristine" and "natural" world has been degraded through human action. They will often invoke a crisis motif to lend a sense of urgency to this loss even while deploying scientific and civic claim-making styles to press home their point (Princen and Finger 1994; Wapner 1996; DeLuca 1999; Chaloupka 2002). What the social constructionist perspective adds to an appreciation of NGOs as moral actors is that it draws attention to the importance of how these actors *articulate* their moral missions (Allahyari 2000). To simply promote a moral society is insufficient, this perspective tells us, because success is affected by the skill with which messages are developed.

Success for many NGOs may also depend on linking moral agendas to perceptions of them as actors that are altruistically inclined. As Sogge (1996b: 13) suggests, "for most people . . . private aid's chief virtues hinge not on politics but on the altruistic gesture." It is when they are seen to fight for the Right and the Good on behalf of others and not simply for themselves that NGOs may actually be best placed to acquire power.

Yet to be seen to be an altruist can be a mixed blessing. There can be difficulties in deciding what the term *altruist* actually describes—giving cause for all manner of heated debate. Further confusing the situation is a general state of affairs whereby "altruistic conduct" is often but a heartbeat away from its alter ego hypocritical behavior—which, among other things, involves "the pretense of virtue, idealism, or sympathetic concern used to further selfish ends" (Grant 1997: 1). Skepticism about human motivation can sometimes mean that the former is identified as the latter—especially, in contexts where powerful elites routinely attempt to pass selfish conduct off in public as other-regarding behavior (J. Scott 1990; Bryant 2002b). More generally, "the pressures toward hypocrisy are immense" in political relationships, where conflicts of interest are common (Grant 1997: 3). Despite these possible obstacles, there is indeed a place for altruism in political processes, but it would seem that "the collective imperative for other-regarding actions must have a compassionate component in order to have any force" (Turner 1993: 506).

Part of the perceived problem here may be linked to the manner in which altruism is often defined in opposition to self-interest. When understood thus, it is not surprising that altruism has come under multidisciplinary assault especially from those rational choice scholars seeking to expunge any explanation for human behavior other than one based on the notion of the completely self-absorbed and utility-maximizing individual (Becker 1981). In social psychology, for instance, some writers have sought to discredit the idea that altruism is an other-regarding practice. In cases where a person acts out of apparent empathy for the suffering of another, it is claimed that the former is only acting to relieve his or her own personal sadness and anxiety. The goal here is to promote psychological comfort, not help another person. This reasoning rules out a role for altruism in human life, dismissing attempts to assert such a role as either misguided or hypocritical (Nagel 1970; Blum 1980; Fultz and

Cialdini 1991; cf. Grant 1997). It asserts instead a universal egoism or psychological hedonism in which the "attainment of personal pleasure is always the goal of human action" (Batson 1991: 5).

Other scholars have come to the defense of altruism. The critique just noted thus fails to recognize that people "never act out of any one overriding motivation but respond to a plurality of varying circumstances," such that self-interest may determine a response on one occasion while altruism might do so on another (Wolfe 1989: 213). It misses, too, circumstances in which people are motivated *simultaneously* by self-interest and altruism—that is, a single response reflecting a mix of impulses. Finally, that critique neglects evidence pointing to a socially constructed sense of the self and other people (Benhabib 1992; Paul et al. 1997).

How, then, to understand altruism? For some, an action is altruistic only if "it is motivated by regard for others" (Schmidtz 1993: 53); others stress that the altruistic act "benefits another rather than the self" (Walster and Piliavin, cited in Bar-Tal et al. 1982: 378). Issues raised here include those of disinterestedness (does the altruist benefit from the action) and intention (is the helper truly motivated to help others).

A more intriguing approach has drawn from work on feminist ethics, thereby bypassing this sort of conventional framing of altruism (e.g., Benhabib 1992; Tronto 1994; Allahyari 2000). Its starting point is the proposition that altruism is an inherently *social* phenomenon. Here, the individual is "de-centered" the better to fully appreciate the merits of altruism. A central assumption is thus that the genesis of motivations, desires, or inclinations is rarely a matter of individuals acting alone. Rather, it reflects a long, complex, and contingent process of social interaction in which an individual's reasons and values are *intersubjectively* determined. As Benhabib (1992: 5) argues, "the human infant becomes a "self," a being capable of speech and action, only by learning to interact in a human community." As individuals develop reasons and values, they inevitably engage in a process of social exchange and meaning construction. A "reason," for instance, is "not just a consideration on which you in fact act, but one on which you are supposed to act; it is not just a motive, but rather a normative claim, exerting authority over other people and yourself at other times. To say that you have a reason is to say something *relational,* something which implies the existence of another, at least another self. It announces that you have a claim on that other, or acknowledges her claim on you. For normative claims are not the claims of a metaphysical world of values upon us: they are claims we make on ourselves and on each other" (Korsgaard 1993: 51).

Once the intersubjectivity of human values and reasons is recognized, there can be a movement away from the view of isolated individuals absorbed purely in self-regarding cost-benefit calculations. This can lead, in turn, to an appreciation of why it is that individuals may want to promote the projects of others. Altruism can be seen thereby to be part of a broader process of human development. The self can be defined in relation to the needs, interests, and perceptions of others even as to grow as an individual is to act in a way that is seen to be socially responsible and beneficial (Wolfe 1989; Benhabib 1992; Smith 1999, 2000).

Indeed, this intertwined process may be integral to individual self-affirmation such that individuals' very sense of moral worth may rest in part on an altruistic disposition. Referring to those who risked their lives to rescue Jews from Nazi persecution, Badhwar (1993: 115) relates that "rescuers had an interest in helping others not just for the sake of others, but also for the sake of being true to themselves and affirming themselves . . . without this necessary connection between their concern for others and their concern for themselves, they could not have loved their neighbor as they loved themselves." Here, what Allahyari (2000) terms "moral selving" is associated with the elaboration of an other-regarding outlook that conditions what individuals do as well as why they do it.

This process whereby moral understanding and altruistic disposition are interwoven may aptly describe a journey of discovery embarked on by many individuals working in NGOs around the world today. Arguably, it is manifested in the missions and visions of organizations, in sector-wide "codes of conduct," in the willingness of employees to accept "substandard" levels of pay and work conditions, and much else besides. It can also be discerned in the sheer passion and emotional energy that is displayed by NGO employees working in an array of political, economic, and cultural settings. If nothing else, the emergence of a vibrant NGO sector has "livened up" politics in all but the most authoritarian of countries today as organizations challenge and sometimes support a wide range of social, economic, and political policies (Fowler 1997; Clarke 1998; Fisher 1998; Meyer 1999; Mercer 2002). In doing so, I argue in this book, organizations may acquire a name for themselves based on the promotion of moral concerns in an altruistic manner with diverse partners.

Context is important here, though, as difference and ambiguity characterize the rise and fall of NGO reputations. The sheer diversity of the "NGO sector" needs to be recalled in recognizing that reputation "profiles" of individual organizations may vary considerably over time and space. Indeed, the relative importance of a reputation may vary between organizations, with some more dependent on it than others. What a reputation for moral conduct may signify to partners of NGOs may differ as well from place to place and from partner to partner, in that notions of morality may be culturally specific, based on"local moralities" (Howell 1997). These moralities often engage, often on unequal terms, with contemporary "Western" moralities that are themselves outcomes of specific cultural and historical "moments" (e.g., Smith 1999; Colloredo-Mansfield 2002).

In short, a cultural geography of morality suggests that there will be variation in what particular peoples in particular places mean by "good" and "bad," "right" or wrong," "just" or "unjust," and so on (Smith 2000).

Yet to acknowledge difference is not to deny the existence and perhaps today even ubiquity of similarity across space and time. Indeed, "conceding the contextuality of moral thinking does not mean that all talk of universals is ruled out" (Smith 1998: 17). For one thing, there is the philosophical argument by the likes of Walzer (1994) and Sack (1997) that, to be human is to share at least a minimal set of basic similarities that amount to a "thin" minimalist but nonetheless important universal moral code (see also Low and Gleeson 1998). For another thing, it is argued by the likes of Corbridge (1993: 463) that the globalization of capital has meant that people's lives are "radically entwined with the lives of different strangers" to an unprecedented degree today, with the result that "there is no logical reason to suppose that moral boundaries should coincide with the boundaries of our everyday community." What Gupta and Ferguson (1997: 37) term "a transnational public sphere" has been part of a process that has "certainly rendered any strictly bounded sense of community or locality obsolete." Further, "the reterritorialization of space" requires us to "reconceptualize fundamentally the politics of community, solidarity, identity and cultural difference" (Gupta and Ferguson 1997: 37).

Many NGOs are part and parcel of this reconfiguration of cultural and moral understanding, as they help to forge new connections between local, national, and global scales. At the same time, they are seemingly able to carve out roles for themselves as moral entrepreneurs moving between actors and scales. "Different" yet "similar" they seem to be, then. Acting in contexts, moreover, that are also "different" yet "similar." These are only some of the paradoxes and ambiguities that condition NGO life (Vaux 2001). There is, too, the matter of precisely how NGO reputations develop, as well as how good reputations may translate into social power.

From Reputation to Social Power

Linking perceptions of moral and altruistic conduct to an ability to act is the question of what it means to acquire a good name. Being in good repute, for instance, with local community groups, donors, state agencies, media organizations, or their peer group seems to be vital to a process whereby NGOs attempt to turn moral vision into action. This is not an easy endeavor. To acquire a good name takes time, energy, and skill as a track record is established. Reputations thereafter "are not inevitable; they may be changed or contested" (A. Fine 2001: 21). Yet, it is when we understand how a multifaceted reputation for moral and altruistic action forms, and how that reputation in turn may facilitate empowerment, that we can appreciate how and why NGOs come to strategize.

Nongovernmental organizations are not unique in sometimes being preoccupied with their reputations. Indeed, reputations are the stuff of everyday gossip and social positioning in relation to a vast array of individuals and groups worldwide (Bailey 1971a; J. Scott 1985, 1990). Yet how significant a reputation is to specific actors can vary greatly. As Scott (1985: 24) notes, for instance, the politics of reputation is "something of a one-sided affair" since "the rich have the social power generally to impose their vision of seemly behavior on the poor, while the poor are rarely in a position to impose their vision on the rich." The same may be true for organizations. Many large states and companies may be less vulnerable to the costs of a bad name than are most NGOs who lack compensatory political-military or economic power. Even here, though, caution is required. For example, business firms spend increasing amounts of money on "public relations" even as they become the object of attacks by anti-globalization protesters and the like (Stauber and Rampton 1995; Klein 2000; *Adbusters* 2001; Richter 2001; Routledge 2003). Still, and broadly speaking, the more an organization can insulate itself from the ramifications of having a bad name, the less likely it will be that it is affected by swings in reputation.

What, then, is a reputation and how is it useful? To begin with then, a reputation is "a socially recognized persona: an organizing principle by which the actions of a person (or an organization that is thought of as a person) can be linked together" (Mercer 1996: 7). It describes a specific quality or set of qualities in relation to an identified individual or organization. Work on reputations suggests a complex social phenomenon. Some of that complexity stems from the fact that a reputation is what others think of you, not what you think of yourself. The self may be "a set of reputations," but these reputations "spring from belonging to a community" (Bailey 1971b: 22). It is ineluctably a relational concept: "my reputation is not something I can keep in my pocket; it is what someone else thinks about me"; as a result, an actor can have "different reputations based on the same behavior" (Mercer 1996: 7).

Reputations do not relate only to individuals. As Gary Fine (2001: 4) notes, we live "in an organizational society and organizations develop reputations that influence their effectiveness." What is more, reputations develop around key words "applicable" to the actor in question. Indeed, several words may be used to shape the reputations of an entire category of actor (such as states, corporations, or NGOs), thereby enabling comparison across organizations. Common criteria for the constitution of reputations should not surprise. Since individual organizations that make up a particular category of actor do similar sorts of things, they tend to prompt clustered perceptions on the part of those who produce reputations. In the case of states, for example, it has been suggested that a reputation in the context of international affairs is notably a matter of "resolve"—a perception of how far this sort of organization will defend its values and interests (Mercer 1996). In the world of business, it is suggested that the reputation of firms is usually linked to questions of transactional honesty and "efficiency" (Klein 1997; Sinclair 2000). In a similar vein, this book suggests that a reputation—as far as NGOs are concerned—is notably associated with the matter of perceived moral and altruistic purpose.

The multifaceted nature of reputations means that individuals or groups may acquire a good or bad name according to different qualities associated with them. For example, a state may be seen to be resolute in confronting international opponents even as a comparable approach in domestic affairs earns it a reputation for being arrogant or callous (Alagappa 1995). Similarly individual business firms may be seen to be honest but not efficient or vice-versa (Klein 1997).

It is perhaps surprising, given such complexity, that reputations are stable at all. Yet they are—suggesting that constructing reputations is a semistructured process. Although reputations vary over time (sometimes sharply), it is nonetheless important to recognize that reputations tend to be *relatively* stable once acquired. This is because they are "collective representations enacted in relationships," such that they are socially embedded (G. Fine 2001: 3). A reputation may generate social consensus even as it can be a focus of dispute.

Central here is an ability to translate a good name into a source of empowerment. What is meant, though, when we speak of "power" in relation to NGOs? The literature rarely acknowledges that NGOs have power at all. Yet there is evidence aplenty to the contrary. Many studies demonstrate the impact of NGOs on social outlooks, donor priorities, business accountability, or policy reform (Clark 1991; Fisher 1993, 1998; Heyzer et al. 1995; Fowler 1997; Meyer 1999; Mercer 2002). True, that impact should not be exaggerated; in fact this book highlights possible limits to NGO influence. Nonetheless, the evidence suggests a type of actor that it is appropriate to describe as *potentially* powerful—albeit, with much depending on specific circumstances (Eccleston and Potter 1996).

Power is commonly used to describe "negative" processes linked to the use of coercion. This need not be the case. After all, power from the Latin *potere* (literally "to be able") simply means to specify a property, capacity, or means to effect things. Recent work emphasizes the diversity of meanings that attach to power. It is what Wittgenstein dubbed a "family resemblance" concept: "just as members of a family resemble each other in a diversity of ways, so, too, the uses of power overlap in a complex interweaving of meanings with no single strand running through the entirety" (Goverde et al. 2000: 18).

One way to understand power is to think about it as having direct and indirect aspects. *Direct power* is the ability to shape *conduct,* and is exercised when A gets B to do something that she or he would not otherwise do. This is "immediate, visible and behavioral, and is manifest in such practices as decision-making, physical and psychological coercion, persuasion and blackmail" (Hay 1997: 51). *Indirect power* is the ability to shape *context,* and is exercised when A can redefine structures, institutions, or organizations such that the parameters of subsequent action by B (and others) is altered. Here, it is "the capacity to redefine structured contexts and is indirect, latent and often [but not always] an unintended consequence" (Hay 1997: 51).

This understanding enables us to widen the scope as to which individuals or groups might be considered to wield power. Thus, beyond politicians, bureaucrats, or business leaders, it is also possible to speak of NGO activists, nurses, social workers, or academics as having power. This latter group is usually not identified in this way. However, their skill at persuasion and negotiation is often such that they too can shape conduct and even context from time to time.

Nonetheless, they usually do not see themselves as possessing power (Edelman 1974; Tronto 1994). As Hugman (1991: 33) notes about the "caring professions": "Individuals or groups who exercise power may be unaware of doing so, and nurses, remedial therapists and social workers may even reject the idea that they exercise power. The lack of apparent conflict in the perceptions of either the powerful or the powerless may be a distortion of the power which provides the basis for their interaction. Therefore, any apparent consensus within a profession, between professions or between professionals and the users of their services must be taken as the object of enquiry as much as any observed conflict." Similarly, those working for NGOs rarely identify themselves as powerful. They are in the business of "empowering others, not themselves"; or, as the leader of one of our case study NGOs put it, NGOs are "facilitators of processes in cultural transmission" (Kalaw 1997: 79). And yet, to help others they need to help themselves first through acquiring adequate power to act.

Thinking of power as sketched above also enables us to appreciate the role of agency in NGO activities. Structural constraints certainly exist here. Yet when structure is exaggerated, there is little scope for speaking of NGO power. Just as strong versions of structuralism do not explain well resistance by the weak to the strong, so too is it true that they rarely capture the potential of NGOs to transform the world around them (Scott 1985; Glassman 2003). Thus, power may permit actors (like NGOs) to shape the conduct of others and (sometimes) even wider social structures at the same time as those structures condition the parameters of the possible (Giddens 1984).

Finally, conceiving of power as above helps distance us from the idea that it can exist only in conflict situations where the strong importune the weak. Indeed, power can be at play in situations where conflict is absent. It can even be a "positive" phenomenon—"power-of" (or "power-to"), not "power-over" (Hay 1997; Doyle

1998). Power can thus be about the social creation of wants and desires in which the generation of shared interests is central. It can be associated with processes of mutuality, persuasion, and negotiation, as opposed to practices of brute force and physical or psychological intimidation.

As the present book illustrates, this understanding of power seems to fit well with the manner in which many NGOs interact with partners. It also helps us to appreciate how an actor may effect change without coercive capacity or economic advantage. This view complements, too, the earlier argument about the intersubjective action between goals and behavior. If people working for NGOs acquire a sense of self in part through other-regarding action, then it stands to reason that power thereby derived is exercised in ways that reflect a sense of mutuality (Jaggar 1983; French 1985).

What is more, power wielded by NGOs, among others, is not to be gainsaid simply because it substitutes the "gentleness" of persuasion for the "harshness" of coercion. Consider the following remark by Lukes (1974: 23): "A may exercise power over B by getting him [*sic*] to do what he does not want to do, but he also exercises power over him by influencing, shaping or determining his very wants. Indeed, is it not the supreme exercise of power to get another or others to have the desires you want them to have—that is, to secure their compliance by controlling their thoughts and desires?" This sort of statement is usually understood as the ability of the strong to convince the weak of the inevitability of the status quo. Yet it can also be applied, albeit shorn of notions of "control" and "compliance," to better understand how many NGOs (and caring professionals in general) can acquire power without necessarily affirming elite interests. It is possible, for example, that power exercised through influence over wants and desires may be directed toward transforming social relations in line with NGO visions.

If power for many NGOs is associated with the art of persuasion and negotiation, then much rides on the perceptions that others acquire of them. The name of the game is acquiring a good name. Reputation forms in the first place only when two conditions are met. First, observers must interpret behavior in dispositional, not situational terms. The reputation producer must explain the reputation holder's actions in terms of that actor's character and not the specific context in which action occurred. Since a reputation is "a judgment about another's character, only dispositional attributions can generate a reputation" (Mercer 1996: 6). Second, observers must use the past to predict behavior in the future. That is, prior conduct is likely to give rise to comparable behavior later on since "people perceive commitments as interdependent or coupled" (Mercer 1996: 6). Reputation is thus important because it reduces social uncertainty and allows trust to develop. Yet a reputation is never static. Since most people usually want a good name, a process of "boosterism" ensues as, in our case, NGO employees talk up their organization's reputation. Here, "impression management" aims to portray NGO motives and behavior in the best possible light in order to increase the likelihood that the organizations will be viewed favorably. However, this is an imperfect and indeed contradictory process. Because reputations are the "gift" of their creator, NGOs can never be sure that their effort will pay off for them.

This uncertainty is compounded by the fact that the production of a reputation is never a neutral process. A reputation is socially recorded in a multiplicity of ways: employment records, donor evaluations, news reports, or even gossip. All of these are inescapably biased. One only has to think about words associated with reputation—*slander, besmirch, laud, censure, character assassination, speak highly of, commend, ridicule, pay tribute to, praise, poke fun at, make light of, scorn, hypocrisy, malign, sully*—to realize that speaking about a reputation is a serious business (J. Scott 1990). It is a business that most NGOs cannot ignore. A good reputation with key partners is seemingly vital to an ability to get things done. Many of those who work for NGOs probably seek to do the "good deed," as they see it, come what may. Yet the argument of this book is notably that they must also be *seen* to do so if they want to be in a position to effect social change in any systematic or long-term way. Still, how *do* many NGOs relate to other actors in a social, political, and economic context not of their own making—and one that can be at odds with the moral vision that got them going in the first place?

Being Entrepreneurial

To think of NGOs as *moral entrepreneurs* is to begin to appreciate some of the many ambiguities that they tend to face. To conceive of them so is to describe a world in which NGOs interact with others *partly* with an eye to accumulating enough *moral capital* to pursue their vision and mission. True, the organization rich in moral capital may not necessarily achieve its goals. Yet the NGO bereft of moral capital is often in an even weaker position to do so. A quest for moral capital thus could be vital to organizational fortunes. To call an NGO a moral entrepreneur is to emphasize such things as resource acquisitiveness, competition, and impression management. Here, this book builds on selected themes in the "dystopian" school and resource mobilization literatures noted earlier. Use of the word *entrepreneur* is particularly suggestive in this regard. That word also evokes a sense of creative risk-taking that may be additionally vital to success. There is an inescapable "liability to risk" in what many NGOs do since the prospect of a reward—broadly understood as an ability to pursue a moral vision and mission—needs to be offset by the perpetual possibility of failure (Skillen 1992). To think about NGOs in this way is not to question the commitment, compassion, and sacrifice shown by many working in these organizations. Nor is it to question the vision and mission guiding many of them. Rather, it is simply to suggest that "good" intentions are more likely to succeed when allied to entrepreneurial conduct. As the next chapter suggests, strategic rationality is the key, as NGOs take risks and link calculation to compassion in pursuit of their chosen agendas (Sogge 1996a; Smillie 1995; Vaux 2001).

CHAPTER 2

The Quest for Moral Capital

One way to appreciate how culture and political economy connect is to describe diverse cultural processes as noneconomic forms of capital. This phrase suggests a world in which individuals and organizations seek to convert noneconomic capital into an economic form (and vice versa). It is also metaphorical, in that it conjures an image of cultural processes shaped by logics of accumulation and consumption commonly associated with the economic domain.

A well-known example of this sort of thinking relates questions of sociability (or social capital) to issues of collective and individual comfort. The emphasis is on social networks and the relative interconnectedness of people. Who people know is important here—not what they know (cultural capital) or own (economic capital). Questions of civic engagement and economic prosperity are linked to such things as interpersonal trust as well as social norms and networks. In effect, social capital has become a kind of benchmark for the ills of modern society in that its perceived absence becomes an indicator of decline. For its leading practitioner, Robert Putnam (2000), even "bowling alone" is read as a symbolic statement about social isolation in America. Such fretting is not new (Putnam 1993; Unger 1998). Nor is the response of critics claiming that Putnam-style analyses ignore the political economy of community development (Portes 1998; B. Fine 2001).

Part of the problem here is that social capital is sometimes hitched to grand narratives about the rise and decline of regions or nations. At this scale, its explanatory value is dubious. The concept may be more useful when related to an understanding of how, why, and with what effect *specific* individuals and groups connect with one another. In basic formulation, it refers to "connections among individuals—social networks and the norms of reciprocity and trustworthiness that arise from them" (Putnam 2000: 19). These networks have "rules of conduct," involve "mutual obligations," and foster "sturdy norms of reciprocity." Through such interaction trust develops—which is important because "trustworthiness lubricates social life" (Putnam 2000: 21). Yet such reciprocity and trustworthiness can promote both social unity and division. Thus "bridging social capital" is inclusive by virtue of involving people across diverse social cleavages, whereas "bonding social capital" is exclusive in that it reinforces social homogeneity. Stated thus, the concept simply says that sociability affects the prospects of individuals and groups. This formulation is not interesting per se. More promising is to reflect on what the ebb and flow of social capital might mean when individuals and groups are seen to be embedded in political and economic structures.

Pierre Bourdieu seeks to address this concern inasmuch as he explores how culture is linked to social stratification. In *Distinction* (1984) and other works, he assesses the ways in which education, taste, and aesthetic outlook shape the reproduction of the more affluent classes. To make sense of such reproduction, Bourdieu develops a multifaceted macro-analytical framework based on the notion of capital. He introduces us to a world in which individuals and groups acquire or lose social capital (social connections), cultural capital (knowledge), symbolic capital (legitimacy), and economic capital (money). For Bourdieu capital is partly a metaphor, inasmuch as it is derived from the language of commerce, but it is also partly a description of actual processes insofar as capital becomes shorthand for "embodied labor" (Bourdieu 1986: 241; see also Swartz 1997).

There are three benefits here. First, the concept of capital provides a detailed and culturally sensitive understanding of power. It encompasses a wider appreciation of the political and economic implications of social interaction than is usual (certainly when compared with the American social capital literature). For example, taste in food is seen both to reflect class-based views "of the body and of the effects of food on the body," and to reinforce those views as it "helps to shape the class body" (Bourdieu 1984: 190). In subtle ways the question of taste is simultaneously a question of class relations, uneven power relations, and finely tuned cultural distinctions.

Second, the framework enables us to recognize how capital accumulated in one area might be translated into a stronger position elsewhere. Fungibility is central here: "The convertibility of the different types of capital is the basis of the strategies aimed at ensuring the reproduction of capital (and the position occupied in social space)" (Bourdieu 1986: 253). True, the degree of fungibility varies depending on the type of capital involved: "The interchange is not equally possible in all directions (Swartz 1997: 80). In particular, "economic capital appears to convert more easily into cultural capital and social capital than vice versa" (Swartz 1997: 80). Yet the act of conversion is important because it provides important clues as to the long-term durability of class-stratified societies.

Third, the framework provides a means to understand the "strategic rationality" underpinning *selected* aspects of contemporary behavior. Multifaceted capital accumulation and consumption occurs when people are aware that benefits can accrue from accumulating different types of capital. Not least, there is recognition that capital accumulated in one area might result later in gains elsewhere (Bourdieu 1986). This suggests that people will seek to arrange their affairs, to some extent at least, in a manner that maximizes utility. As such, there seems to be a process of instrumental reasoning and flexible cost-benefit calculation going on here somewhat akin to the more sophisticated variants of rational choice theory produced in recent years (e.g., Bridge 2000; Chong 2000).[1]

With Bourdieu, this is arguably a process that opens up the sorts of behavior to be "costed," the time frame over which costing is seen to apply, and the nature of the benefits that might thereby be enjoyed. True, Bourdieu suggests that his theory of human action is broadly about "tacit" and "unconscious" strategizing by actors (e.g., Bourdieu and Wacquant 1992). He clarifies why, for example, the upper class devotes time and money to the pursuit of cultural refinement via prolonged formal education or the cultivation of social connections through club memberships instead of focusing on evident political or economic pursuits. Still, it is not clear how tacit and unconscious such strategizing really is; at a minimum, Bourdieu is ambiguous here since he does "recognize degrees of awareness of the interested character of some forms of action" (Swartz 1997: 70). As such, the association of Bourdieu with a sophisticated utilitarianism is not far-fetched after all.

Bourdieu's multifaceted "capital" approach has nonetheless been the focus of criticism. One concern is how use of this metaphor allows "inappropriate" application of economic logic to noneconomic activities. Sayer (2001: 176), for instance, believes that Bourdieu's "cultural turn with an economic twist" conflates use value and exchange value. Sayer (2001: 178) believes that, because Bourdieu does not distinguish between the two, the inevitable outcome is that he "reduces the former to the latter, emphasizing investment, calculation and profit." As a result, in Bourdieu's framework there is no allowance that what seem to be debates over judgments of taste might actually be debates over intrinsic value—and not instrumental value linked to competitive advantage for selected participants (B. Fine 2001).

There is also criticism more generally over the commensuration of values in much rationalist thinking. Commensuration creates "an abstract form of unity that is capable of encompassing any valued thing . . . [It thereby] creates relationships among extraordinarily diverse and remote things" (Espeland 1998: 28). Thus it enables tradeoffs and comparisons—and ultimately decisions—even in contexts where values sharply diverge. This critique, particularly its insistence that some values are incommensurable, is a useful attempt to highlight the excesses and distortions of rational choice theory, especially its most strident proponents (e.g., Becker 1981). Espeland (1998), among others, demonstrates the plurality of values and rationalities that can make a mockery of cost-benefit decision-making processes (see also Radin 1996; Zelizer 1997). At the same time, strict adherence to value incommensurability is itself "a murky subject" (Radin 1996: 9)—not least, it would seem, because of the potential paralysis that might be induced in much of contemporary social life if it were ever the basis of social and political action. The incommensurability view also returns us to the particularism versus universalism debate alluded to earlier, inasmuch as, if taken to an extreme, it could lead to an ethical "relativism" requiring "equal respect for all moral codes [values], including those incorporating human sacrifice, institutionalized torture and so on" (Smith 2000: 15; see also Benhabib 1992).

These concerns, which relate directly or indirectly to the capital framework under discussion here, can be usefully situated too in the wider context of debate over communicative and strategic rationality. The capital framework is a way to understand some of the implications of the historic spread of the sort of strategic rationality considered in this book—instrumental thinking often (but not always) linked to the assertion of exchange value in an increasingly neoliberal world (Wolfe 1989; Johnston et al. 2002). In contrast, Habermas's (1984) notion of communicative rationality posits a broader understanding of rationality—one in which communicative action is notably a quest for mutual understanding, not strategic manipulation. In the Habermasian view, strategic rationality of an instrumental kind is central to behavior in the realm of material production and reproduction (the "system"), whereas communicative action is pivotal in the "life-world" or realm of cultural traditions, social connections, and normative processes.

Yet Habermas and others concede that a key trend in recent times has been the progressive colonization of the life-world by systemic processes. As a result, an increasing proportion of human activity is shaped by strategic rather than communicative action. At issue is the "expansionary logic of capitalism," which is predicated on the substitution of exchange value for use value and strategic action for communicative action (Miller 1992: 34). It is in this regard that a capital framework resonates since it strives precisely to understand the strategic and often instrumental basis of what increasingly passes for contemporary culture in many parts of the world today. Clearly, "cultural globalization" is a multifaceted, contingent, and hybrid affair overall (Gupta and Ferguson 1997; Ong 1999; Johnston et al. 2002). Yet, if "difference" is not thereby erased, this historic and worldwide (and ongoing) transformation nonetheless points to, on the one hand, the growing futility of thinking about culture as a separate, neatly bounded "thing," and on the other, the increased utility of thinking about transnational cultural logics of social and political practice. One such logic (or set of logics) arguably is related to those NGOs (and their partners) that operate at multiple scales and connect cross-culturally and cross-nationally (as with the NGOs examined in this book; see also Fisher 1993, 1998; Wapner 1996).

That strategic rationality with strong utilitarian overtones is seen to have a growing purchase on human life today is reason alone to take seriously the capitals perspective. Another reason relates to the importance of understanding "strategic rationality" itself. As Bridge (2000: 528) notes, this idea is not synonymous with "*homo economicus*—that pared-down, mean spirited, selfish actor" much beloved by some proponents of rational choice. Rather, strategic rationality is about the view that "agents will act consistently in the pursuit of goals that are consequential to them and that will involve taking into consideration the plans and purposes of others." There is certainly an assumption here that people will generally seek to act in their own "self-interest"—a view that clearly runs through much work by Bourdieu (Swartz 1997; Bridge 2001). However, such an interest "can encompass any human motivation (from altruism to egotism)" (Bridge 2000: 528). Further, that interest can shift over time as a result of interaction with others and/or personal reflection—that is, it is socially constructed. In short, strategic rationality—including its instrumental variant—may combine calculation with a flexible vision of means and ends that is often appropriate for actors (such as NGOs) embedded in often complex webs of relationships.

Thus understood, the notion of strategic rationality emphasizes the colonization of the life-world by systemic processes linked to an advancing if heterogeneous global capitalism. Unlike much rational choice thinking, though, it provides ample scope for interpreting a wide range of social practices that may be separate from or even resistant to the triumph of *homo economicus*. Thus an investigation of strategic rationality broadly in the manner of Bourdieu need not, as some allege, result in an inability to think critically (B. Fine 2001; Sayer 2001). To the contrary, the use of this concept enables a comprehensive understanding of the complexities and ambiguities of social control, acquiescence, opportunism, and resistance in the modern era (Swartz 1997).

Indeed, the approach sketched above may have much to offer in understanding the situation of NGOs operating in the neoliberal era. A multicapital frame of reference guided by a strategic rationality with an instrumental flavor seems broadly appropriate when considering the particular development history and social position of many NGOs—although such a claim must be assessed in relation to the actual experiences of organizations shaped by diverse conditions and beliefs. Clearly, a book of this kind can only begin the task of assessment but can be informative nonetheless in suggesting possible areas of empirical utility in relation to specific organizational traits and contextual conditions.

Yet an approach that broadly privileges a role for strategic rationality and noneconomic capital as set out above must be mindful of the hazards of conceptual stretch. This is clearly a problem with the U.S. social capital literature (e.g., Portes 1998). It is also a difficulty, though, that is sometimes attributed to the work of Bourdieu. Thus, for example, Dreyfus and Rabinow (1999: 91) observe that "everything from accumulating monetary capital to praise for being burned at the stake automatically counts as symbolic capital" (see also Swartz 1997).

What is needed is a more focused approach tailored to specific circumstances or actors. This is an approach that aims for micro-analytical depth not macro-analytical scope. The following discussion develops such an approach around the notion of moral capital (see also Kane 2001). It does so with an eye to assessing selected aspects of NGO strategizing. Here, questions of reputation and strategic rationality of an instrumental hue come together as it becomes possible to think about NGOs as moral entrepreneurs. To the extent that organizations seek moral capital, they are involved in a process of reflexive behavior that has tangible consequences, in terms of political, financial, and territorial strategies.

Moral Capital and Empowerment

I understand the quest for moral capital as a purposive endeavor that is concerned to build an organizational reputation for moral and altruistic action. I am thus not concerned per se with the pursuit of moral capital either inside or outside of NGOs by individuals—although it is acknowledged at various stages that these two processes can be linked. Instead, this study is concerned with the fates of organizations, which can be seen to have trajectories all their own apart from the individuals who have worked for them.

If moral capital is akin to a reputation for moral and altruistic action, then the concept encapsulates many of the arguments about morality, altruism, reputation, and power examined in the last chapter. That it is, like reputation, a socially constructed "good" dependent on the perceptions of others is vital, since much of its dynamism—and perhaps even the fate of organizations that depend on it—is an outcome of this situation. Yet use of the term *capital* is advantageous precisely because it serves to emphasize additional factors crucial to a rounded understanding of NGO reputations. To begin with, it draws attention to the possible practical *utility* of a reputation for an individual or an organization. For NGOs, a reputation for moral and altruistic action is important not so much because it might be inherently valuable but rather because it helps create opportunities otherwise unavailable. It can be a source of NGO empowerment.

The term *capital* also conveys a sense of *agency* appropriate to a study concerned with showing how organizations may boost their reputations through reflexive action. True, a reputation is something that others think about you. Yet that does not preclude intervention aiming to shape what significant others think about you. Moral capital can thus put a premium on the agency of the *receiver*—as well as the producer—in a way that opens up a world of strategic thinking that is the subject of this book. Here, *capital* evokes strategic rationality of an instrumental kind inasmuch as it is about "action" or behavior "for the sake of tangible, exterior returns" (Kane 2001: 7). Indeed, the term serves to introduce an element of *contingency* and *ambiguity* that positively demands that organizations behave strategically. Moral capital, as with other sorts of capital, "requires both continuous skill and luck in its maintenance and deployment" (Kane 2001: 7). True, a reputation seems to involve a measure of stability as a collective image is generated around specific individuals and organizations. Yet stability must not be exaggerated since consensus about an individual or organization is never guaranteed. Clearly reputation is indisputably political and contingent, as it is associated with the interests and prejudices of reputation producers (Bourdieu 1986). All of which places NGOs in a quandary as to how to behave if they are to pursue moral capital with diverse partners concurrently. As will be seen, this dilemma helps to explain ambiguities in NGO action.

The quest for moral capital is thus about how cultural resourcefulness may be expressed through an instrumental kind of strategic rationality. It is about assessing what it means to think about this sort of organization as a moral entrepreneur. Words such as *resourcefulness* and *entrepreneur* are not used lightly (Schoenberger 1998). They emphasize how a quest for moral capital can be enabling even if outcomes are partly dependent on what organizations do themselves. I return to the question of moral capital and strategic behavior below since the main concern of this book is to understand how the quest for the former can be reflected in the latter. Here, I consider some of the possible *benefits* that accrue to NGOs with moral capital in order to relate the process to their empowerment—and thus, why organizations might pursue that quest in the first place.

To begin with, moral capital may help to strengthen an organization internally, inasmuch as it contributes to self-esteem and solidarity among those who work for it. Indeed, a virtuous circle may develop. Employees earn the respect of others for effective deeds. This can, in turn, reinforce both individual self-esteem and collective pride, as work by the organization is valued. A response here can be greater motivation, perseverance, and optimism about the NGO's impact than might otherwise be the case. Thus an employee at one organization featured in this study attributed low staff turnover to the fact that they "loved the PAFID [Philippine Association for Intercultural Development]." A worker at our other case study, the Haribon Foundation, explained his attachment to the organization as a "vocation"—"I love my work," he declared. True, these sentiments are not expressed by everyone, and some employees certainly move between organizations during their careers. Yet benefits are genuine when a virtuous circle occurs. Thus "solidaristic motivations stem from powerful emotional drives and generate their own significant rewards"—thereby helping to explain the "staying power of many agencies with very limited resources" (Brett 1993: 283).

Even more remarkable are benefits associated with changes to an organization's *external* working environment. Here, NGO empowerment finds expression inasmuch as moral capital is used to promote altered practices, knowledge-claims, and ethics on the part of others. This is not an automatic process. As Kane (2001: 7–8) observes, "It is not enough to be good, or morally irreproachable, or filled with good intentions, or highly and widely respected. It is necessary to have the political ability to turn moral capital to effective use, and to deploy it in strategic conjunction with those other resources at one's disposal . . . [I]t may be well or foolishly, fortunately or unfortunately invested, it may bring large returns to oneself or one's enterprise or it may be wasted and dissipated . . . [and] there are always opponents with a vested interest in doing everything they can to ensure dissipation." The consumption of moral capital requires skill and strategic ingenuity just as in its accumulation. As I show, the accumulation and use of moral capital are not unrelated. Thus, even though our focus is on accumulation, the discussion at times also assesses the use of existing moral capital.

Consider the altered practices that might result from skilled use of moral capital. Many of them are a basis for action. Thus the ability to raise funds for an organization's work can be a major benefit linked to moral capital. This link is usually strong, if never guaranteed. Moral capital may indicate that an NGO can be trusted to spend funds wisely. A donor may also conclude that it will gain prestige itself from supporting the work of an organization. As a senior Ford Foundation officer put it, her organization sought to "back innovative people" who would make a "significant difference" with the limited funding that the agency could provide. Indeed, NGO "rootedness" in local communities was deemed vital if support was to be forthcoming (Racelis 1997; see also Ford Foundation 1997).

Moral capital can facilitate spatial and territorial strategizing inasmuch as it can help locate and retain partners at the local level. While the moral vision and mission of NGOs is often linked to promoting social justice, this agenda never guarantees the support of local residents in areas where organizations wish to work. Moral capital can serve as an antidote to the skepticism and distrust that often confronts NGOs in these situations. Area-based reputations seem especially important in this regard. For example, the PAFID's strong reputation in northern Luzon led residents in one community into "appealing before you our urgent problems that are very pressing and for which we need your utmost help and assistance nowadays" (Residents of Mapayao 1992). Moral capital may also translate into a measure of influence over state policies and practices. Such influence is rarely as much as desired, and it is usually associated with "small victories"—the defeat of a proposed cement plant here or the award of an ancestral domain certificate there, for example. Still, moral capital may lead to valuable intelligence, technical support, or funding from state agents sometimes fearful of political difficulty in the absence of action. It may mean that an NGO can offer criticism without forgoing influence. The records of the Haribon Foundation in a national anti-logging campaign and the PAFID in a donor-funded upland project provide contrasting evidence as documented below. What can be said here is that anti-state agitation is not necessarily incompatible with having good access to official circles—at least for the reform-minded organizations featured in this study. Indeed, the two are linked in a complex way that puts a premium on multifaceted political action by NGOs keen for change.

Media exposure can also be linked to moral capital. In looking for a likely interviewee, there is a tendency among journalists to go with individuals and organizations known to have a good "track record." This was the case for the Haribon Foundation during the anti-logging campaign, when it all but monopolized media coverage, thus becoming widely known in the country. Here again, though, evidence will show that the effect of using moral capital can be ambiguous, as a reputation is boosted in some quarters while being deflated elsewhere (DeLuca 1999; Smith 1999).

An NGO with moral capital may win public support. This can facilitate fund-raising, since private citizens usually donate money to organizations with which they are familiar *and* which they believe will use the donation effectively. There may also be support in the form of voluntary labor providing diverse benefits: consumer boycotts, protest marches, work in thrift shops, and so on. The ability to pursue a vision may thereby be enhanced. The volunteer route is "soulful" as well as providing tangible benefits to an organization (Fowler 2000; Allahyari 2000).

These altered practices illustrate how moral capital may be a basis for NGO action. They involve altered behavior by others: decisions about funding, access to official information, community receptivity, media coverage, voluntary labor, and so on. These changes can seem quite modest, yet in aggregate their attainment may be critical to empowerment. As such, strategic behavior associated with the quest for moral capital analyzed herein revolves around these "bread-and-butter" issues.

Still, there is more to the question of moral capital and NGO empowerment than such incremental change. Indeed, there may be fundamental transformations afoot in the way people think about issues. Here, it pays to remember that many NGOs promote social change—be it reformist or radical—and hence, do not operate simply to promote themselves for the sake of it (Fowler 2000). The two processes are linked, inasmuch as the latter may be needed in order to advance the former. Yet a bigger agenda means organizations may be keen to promote new ways of thinking and caring consonant with their vision and mission (which are, themselves, open to change over time). Success here, though, seems to be conditioned partly by organizational moral capital. A good reputation skillfully deployed may prompt the acceptance of new knowledge and ethical claims. Social practices may then change as ideas are acted on in line with NGO preference (e.g., Princen and Finger 1994; Fisher 1998; Bryant 2002a).

This process involves organizations in the elaboration of moral discourses. These discourses come in many shapes and sizes, yet they all seek to specify a social or environmental "problem" even while justifying action by an NGO in its resolution (Forsyth 2003). This is precisely what the vision and the mission of most NGOs are all about. Consider how moral discourse is constituted. As Ibarra and Kitsuse (1993: 35) explain, rhetorical idioms function as "moral vocabularies" which provide those involved with "sets or clusters of themes or 'sacred symbols' capable of endowing claims with significance." Rhetorical idioms structure claims and perceptions along certain lines. They also "locate and account for the claimant's participation in the social problems language game by reference to moral competence (instead of strict self-interest)," so as to enlist another to make either "sympathetic moves" in that game or sustain such moves "by the already converted" (Ibarra and Kitsuse 1993: 36). Specific rhetorical idioms proposed by claimants (in our case NGOs) include rhetoric of loss, entitlement, endangerment, or calamity. Rhetoric of loss, for example, is concerned with inveighing against the devaluation of some sacred or highly valued object, such as a "pristine" natural environment. It invokes the concept of protection as a remedial course of action in which the claimant may play a leading role. In contrast, a rhetoric of calamity aims to shock participants by using metaphors and reasoning practices that "evoke the unimaginability of utter disaster" and the need for remedial action without delay (Ibarra and Kitsuse 1993: 41; see also Hannigan 1995).

Rhetorical idioms are expressed with reference to motifs that are "recurrent thematic elements and figures of speech that encapsulate or highlight some aspect of a social problem" (Ibarra and Kitsuse 1993: 47). Such motifs may combine a sense of urgency with moral judgment through words like *crisis, scourge,* or *menace.* These motifs, and the rhetorical idiom of which they form a part, are presented through claims-making styles in which claimants assert claims through particular styles of presentation. These styles include the scientific (disinterested, technical, objective), the comic (absurdities, irony), the theatrical (issue dramatization), the civic (honest, unpolished moral outrage), and the legal (representation of a defendant, justice).

The point about these moral discourses and associated styles is that they are intimately associated with moral capital relationships. The production of moral capital is certainly linked to assessments of individual NGOs. However, this process does not stand apart from moral discourse but is connected to it in complex ways (Martin 2003). In some cases producers may even come to view things in a way akin to that of its NGO partner. For example, many indigenous communities working with the PAFID have overcome initial skepticism of official land tenure agreements. They have come to view them as suitable interim measures on the road to full recognition of ancestral domain. For their part, donors acknowledge how NGOs such as the Haribon Foundation or the PAFID have "formulated the issues" by playing a "very important role" in a context of ineffectual government (Braza 1997; Chua 1997; Racelis 1997). In short, moral capital may help an organization to pursue long-term change via its partners.

Long-term influence is most likely, though, when moral capital is used to persuade others of the need for a new ethics. That altered ethical views may emerge from moral capital relationships is not surprising. After all, if NGOs build reputations for moral and altruistic action, then it stands to reason that they may be well placed to influence those around them. There can be far-reaching implications when new ways of caring are disseminated (Smith 2000). As Kane (2001: 256) suggests, "what once we unthinkingly regarded as lacking moral connotation has proved to be deeply implicated in matters of justice, and innocence lost can never be regained." In the process, an obligation inspired by ethical reflection can lead partners into the recognition that they not only *ought* to behave differently, but that they *must* do so to remain true to new-found identities.

They may even come to believe that others must behave in a new fashion, thereby widening further the purchase of the NGO vision inasmuch as that vision is seen by others as requiring collective transformation as well as individual change. This is because morality "draws implications for judgment and action from the emotions and cognitive understandings that people hold" (Jasper 1997: 135; see also Snow and McAdam 2000).

The ability to use moral capital to effect change in people's practices, knowledge, or ethics is highly dependent on the political skills of organizations. In particular, the articulation of moral discourses involves a process of "frame alignment" whereby the cognitive frames of NGO employees and potential associates are aligned in terms of problem definition and solution as well as appropriate motivation. At issue is the generation of a collective action frame, understood as an "interpretative schema that simplifies and condenses the 'world out there' by selectively punctuating and encoding objects, situations, events, experiences, and sequences of actions within one's present or past environment" (Snow and Benford 1992: 137; see also Snow et al. 1986; Jasper 1997). This process can be thought of in terms of strategic rationality, as a means by which an individual or group appeals to others for support (Martin 2003).

The concept of frame alignment has been used mainly to appreciate the strategies used by social movement organizers to win recruits. Recently, it has been extended to encompass how the personal identities of recruits are aligned with the collective identity of the movement (Snow and McAdam 2000). In both cases, though, attention is focused on framing as an *intra*-movement phenomenon. Meanwhile, work in rhetoric studies sees framing as a political tool of protest. Thus DeLuca (1999) considers the "image events" used by groups such as Earth First! or Greenpeace as a means to "televisually disrupt" predominant understandings of the environment. Yet consideration of moral capital relationships suggests a different take: as an outward-looking strategy of an NGO to recruit support from partners to aid its aims.

If the quest for moral capital is to succeed, NGOs must become "master framers." This is rarely straightforward, however. It is a function of the properties of moral capital and reflects the complexities that result from the dense network of actors in which NGOs tend to find themselves (Keck and Sikkink 1998). What is clear is that the accumulation of moral capital provides a basis for subsequent use, resulting in important benefits for an organization. There is, in short, ample reason to think like a moral entrepreneur. Still, what does this mean in practice and how might action be conditioned by it?

Moral Capital and Strategic Behavior

Nongovernmental organizations need strategies to acquire moral capital, since acquisition (let alone retention) requires skill. Moral capital is not a birthright of "charitable" organizations but must be earned and used wisely to be an effective resource. This task is quite daunting since NGOs must persuade others to create moral capital for them. Indeed, due to the fissiparous nature of the process, accumulation with one partner might even lead to dissipation of capital elsewhere. True, the situation is not often a zero-sum game. Still, NGOs can be tugged in different directions by the expectations of partners. Consummate strategizing is then key here, especially when an organization must navigate between incommensurable aims.

As such, strategic thinking and action—far from undermining the moral values and altruistic concerns of an organization—is often vital to communicating values and concerns. In this regard, Chong (2000: 227) suggests a need to distinguish "unconditional moral or expressive behavior and moral or expressive behavior that is coordinated with the choices of others." When the latter is present, actions "can typically be understood only if due consideration is given to more systematic strategic processes"—behavior moreover "generated and encouraged by social coordination and community pressures" (Chong 2000: 227–28). Strategic rationality with an instrumental tone is certainly not the whole story, as noted earlier. Yet it helps us to understand how NGOs might acquire moral capital in ways that enable moral visions and missions.

To some extent, this process encourages institutional boosterism as NGOs seek to "manage" their public images. Such action is not on a scale comparable to what we see in the case of large corporations or political parties. Here, public relations professionals are deployed to "shape organizational images" or to massage leaders' images so as to promote more favorable conditions (A. Fine 2001: 5; see also Stauber and Rampton 1995; Richter 2001). Still, NGO impression management is on the rise, led by large Western organizations desirous of a professional and "caring" image. The "slick image" of American NGOs has yet to be acquired by Philippine counterparts, but there is a move in that direction (Racelis 1997). One observer remarked that Philippine NGOs are ever more concerned about putting "their best foot forward" as they "put a premium" on having "a good name" (Braza 1997). Another informant noted that there was "definitely a change" when it came to "image building" with many organizations keen to "advertise themselves" through media work and public events (Chua 1997).

It is unlikely, though, that Philippine NGOs will end up resembling some packaged product emanating from Madison Avenue. It is one thing to promote "good deeds" to partners and quite another to make deeds up—even if such things as "glossy" pictures in brochures sometimes push the limit here (e.g., Conservation International 1995). Exaggerated or false claims may be discovered as partners swap accounts. Indeed, even the cavalier treatment of weaker partners, such as local community groups, can have a debilitating effect on an NGO (Bryant 2002b). Once exposed, such practices may dog the perpetrator because such deceit is so at odds with expected conduct. Institutional character is sullied as a reputation for deceit and hypocrisy grows (Johnson 1993; Grant 1997). The benefits of moral capital are withdrawn. Thus, for example, donors, state agencies, or local communities may blacklist an organization. The vicious circle associated with dissipating moral capital is considered below. The point here is that there may be "limits" to institutional boosterism quite aside from constraints imposed by a lack of resources. Impression management thus tends to be "low tech" inasmuch as it revolves around media work, conference presentations, strategic briefings, or newsletters full of "success stories"—even as these activities are often multipurpose in that they also serve other ends such as political mobilization.

Still, many Philippine NGOs feel pressed to dedicate scarce resources to promoting their image. There is little room for reticence here in a competitive sector (Alegre 1997; Fowler 2000). If an organization relied exclusively on word of mouth to build a reputation, moral capital would tend to expand slowly and erratically due to reliance on a "slow" form of communication. There would be no guarantee either that news of a good name would reach all of those that the organization wished to impress. I do not denigrate the potential importance of word of mouth here. In fact, chapter 6 highlights how the development of area-based reputations is partly a matter of precisely this sort of practice. Yet, it often requires a supplementary strategy of targeted publicity. That others may be boosting their own names is another incentive. Even a reticent NGO is likely to react when it is felt that it might lose political access, economic support, or media attention. As a case study NGO lamented: "Those who can speak and write more convincingly and appear more handsomely in front of everyone are getting a big slice of the donor's cake. The PAFID, long known for its silent but very relevant job, was left with no choice but to join the bandwagon" (PAFID 1995a: 1).

Yet this concern should not be exaggerated. The quest for moral capital involves much more than PR. It requires multifaceted initiative, notably encompassing political, financial, and territorial strategies. These strategies need flexibility to account for changes in the work environment. They can also be ambiguous as organizations struggle with trade-offs. What these strategies have in common is sensitivity on the need to build bridges between people, to develop common agendas, and to nurture mutual respect.

That sensitivity, strategic rationality, and moral capital become intertwined is seen when NGOs pursue multifaceted action. There is a tendency to promote area-based reputations with attendant spatial economies of scale often facilitating acquisition of moral capital. It is seen too in the way that organizations would define the terms of political engagement. As NGOs promote constructive engagement with local community partners and critical engagement with state partners, tradeoffs in conduct often reflect sensitivity to what various partners think is appropriate. For instance, an organization may criticize a specific policy out of belief. Yet it may also do so because of public *expectation* that such criticism will be made when policy hinders promotion of a mission. As one journalist noted, NGOs are seen as "public watchdogs"—in the business of "being vigilant" vis-à-vis the powerful (Vitug 1997). Finally, strategizing is seen in the way that organizations promote financial well-being. Although autonomy-boosting measures involve public fund-raising and creating a diversified donor portfolio, financial strategies are not reducible to a straightforward dash for cash in a context in which moral capital is also prized (Fowler 2000).

Moral capital is integral to organizational practice. It can influence strategy as NGOs select where to work, from whom they want funds, and with whom they will engage (and on what terms). Yet these decisions are further complicated by the context in which moral capital operates. Strategic choices reflect a wider social context. An NGO may seek moral capital but does so in the knowledge that others are doing likewise. The pursuit of moral capital is a process in which organizations interrelate as allies and/or competitors in furthering their agendas.

Moral capital thus seems to have a contradictory effect on inter-NGO relations. Conflict is one outcome, as competition is manifested through funding battles and "turf wars." Solidarity is not inevitable when NGOs behave like moral entrepreneurs. Yet the conclusion that organizations are caught up in a perpetual "war of all against all" is also false. Cooperation too is an outcome of the workings of moral capital. NGO coalitions are one example, but it may also be seen in quotidian practices ranging from the exchange of information to turf etiquette. This may simply reflect a wish for solidarity, of course (Brett 1993). However, it may also be recognition that the pursuit of an agenda can best be achieved through collaboration. When moral capital is accounted for, an ambiguous picture emerges of NGO strategic interaction.

Moral capital is most evident at times of organizational crisis. As Kane (2001: 11) observes, "its crucial supportive role is not clearly seen until it is lost and individuals or institutions face consequent crises of legitimacy and political survival." If this is so for the political leaders that Kane writes about, it is also true about the NGOs examined in the present study. Organizations will go to great lengths to reclaim or defend moral capital. They will defy powerful opponents and snub powerful allies if it is felt that unacceptable action is being required of them. Sometimes decisions are made that defy narrow economic logic—for example, when funding is rejected even though resources are scarce. Sometimes decisions are made that put staff in danger, as when work continues in an area despite threats of violence. Here, the link between moral capital and strategizing often stands out.

In some cases the loss of moral capital may reflect ineptitude. An organization may hemorrhage capital as one partner after another becomes disenchanted. Or sheer bad luck may be to blame, as when an NGO is in the wrong place at the wrong time. It may even be a combination of factors. Further, there may be no sharp crisis to serve as a warning but simply deflation as moral capital ebbs away and staff members are less able to promote their agenda. The result may be the moral capital equivalent of bankruptcy. This book does not focus per se on hapless organizations of this sort. That said, it does offer evidence of a loss of moral capital by the two featured NGOs—thereby illustrating how even "successful" entities can make "mistakes."

Moral capital thus seems to be part of the fabric of NGO life, for better or for worse. Its presence can make or break an organization. It would be wrong, though, to attribute too much influence to this seemingly important if little studied resource. What NGOs do also reflects other influences. Thus action may be expressive inasmuch as it is "performed for its own sake, with no apparent rational consideration of material consequences for the actor" (Chong 2000: 223). It may reflect social networks (social capital) where it is who works where—not reputation—that matters most (Lin 2001). It is likely, though, that social and moral capital go together. There is often an overlap between whom you know and whom you respect. The potential relationship is illustrated at various stages below. Yet they are not identical. Meanwhile, action may also suggest that organizations are rich in socially scarce knowledge (human capital). What employees know, not what others think about them, is crucial (Eyerman and Jamison 1991). Once again, though, the situation is rarely either/or. Moral capital may coexist with a respect for possessing sought-after skills.

Then there is the view that NGO action simply reflects funding opportunities. I clearly do not subscribe to this view. True, funding is a source of much concern and anxiety. Yet this does not mean that organizations will do anything for money. The evidence presented here, and elsewhere, is unequivocal: organizations devote much effort to looking for ways to promote financial autonomy (e.g., Bennett and Gibbs 1996; Alegre 1997; Fowler 2000). This effort would not make sense in a world where pure economic gain was the sole factor. Indeed, considerations linked to moral capital can play a vital if still little understood role in the moderation of economic imperatives. Moral shame (let alone political ineffectiveness) may be a fate worse than poverty. Many organizations lead a hand-to-mouth existence in any event (Nozawa 1996).

The moral capital approach set out above offers a novel perspective on NGO life, making the case for the likely importance of non-economic capital in general and moral capital in particular. Yet how might the quest for moral capital be studied empirically?

Investigating Moral Capital

Moral capital would seem to be potentially important in shaping what many NGOs do as well as how they come to influence others. It has a bearing on organizational strategizing and empowerment. If accumulation and use of moral capital are linked, it is nevertheless useful to distinguish them analytically. The former is about orienting action to promote a good name; the latter is about what is attempted once a good name is achieved. My main concern in this study is with the accumulation of moral capital. This is partly out of a pragmatic wish to keep the work to a manageable length. Yet it is also partly out of a sense that it is on the accumulation side of things that greatest insight will be gained into the strategizing behavior of NGOs—and hence, the best opportunity to gauge the utility of the theoretical framework set out in this chapter.

The focus is then squarely on NGO strategic behavior. Since I am interested in "the intentional manipulation of cultural resources" to further ends, this focus is apt insofar as a "strategy" is a popular method by which to do so (Chong 2000: 229). Indeed, this focus is consonant with my sense that a moral capital approach usefully emphasizes human agency in the acquisition of moral capital. And skill is required here. To acquire moral capital is not only to "do good," but above all to be *seen* by others to be "doing good." This requires a proactive and strategic response by an NGO. And yet, it needs to be reiterated that to be seen to be *too* keen in the quest for moral capital can often be self-defeating as partners may become skeptical of an organization's intentions (Hulme and Edwards 1997). There is an abiding tension here in strategic action that NGOs need to be cognizant of if they are to avoid an unintended and damaging loss of moral capital.

I am concerned with what an organization does to promote a reputation for moral and altruistic action rather than what specific partners actually think about it. The evidence is thus based mainly on the words and deeds of the two Philippine NGOs analyzed: the Haribon Foundation and the Philippine Association for Intercultural Development (PAFID). Additional evidence comes from other sources insofar as it sheds light on possible links between moral capital and strategizing. There is no effort therefore to "judge" the two organizations. Rather, their role serves a different purpose in that they are somewhat akin to "case studies of moral capital in action" (Kane 2001: 9).

The case study methodology is appropriate to this inquiry because it is ideally suited to the testing of theoretical propositions—viz., the importance of moral capital to NGO strategizing (notably political, financial, and territorial aspects). As such, the two NGOs were *not* selected because they were "representative" but because they were suitable for an investigation of the theoretical concerns. Both organizations are well established and work in many areas of the country. They are mid-sized entities (between thirty and fifty paid employees) by Philippine standards; as such, they have sufficient complexity to permit the elaboration of the multifaceted strategic behavior of interest here. Further, their records were of sufficient duration and richness—involving routine and crisis-related actions, for example—that there was a good prospect that the role and impact of moral capital would be readily discernible. At the same time, their diversity—Haribon is an "environmental" NGO whereas the PAFID is a "development" NGO—provided an opportunity to assess the possible importance of moral capital in relation to strategic behavior across two sectors. Yet the point was not to generalize findings to the NGO population as a whole but to relate the findings to starting propositions. This is therefore an exercise in theory building, not population description.

Case study research is appropriate in situations, like this one, in which the aim is to seek causal explanation rather than descriptive accuracy. It is helpful where the aim is to look for evidence in relation to the questions "why" and "how" rather than "what." The case study also enables research in a context characterized by multiple sources of evidence, and in which setting explanation is derived from data triangulation (rather than data repetition, as in surveys). Finally, case study research enables the analysis of situations in which there are "many more variables of interest than data points" (Yin 1994: 13; see also Hakim 1987; Hamel 1993).

In this study, data triangulation was obtained on the basis of three main sources of evidence. There was, first, extensive analysis of documents by and pertaining to the NGOs. Evidence was gleaned from NGO reports, pamphlets, newsletters, minutes of meetings (including board meetings), staff field reports, and correspondence.

In the case of both NGOs, this process enabled an overview of various aspects of NGO operations including finance, administration, employee relations, and external relations. This material was then supplemented by evidence taken from reports by NGO partners such as donors and state agencies as well as from newspaper and scholarly accounts. Yet there are potential problems with this source of evidence in general. For example, there may be biases in the selection of documents, including the denial of access to confidential and potentially unflattering material.

Hence, a second source of evidence was used in the form of in-depth interviews. Nearly one hundred interviews were conducted, with a typical interview lasting about two hours (a complete list is given in the bibliography). One set of interviews encompassed a wide range of individuals based in state agencies, Congress, donor agencies, academe, businesses, and media outfits, with the aim toward gaining an overview of NGO strategizing. The other set of interviews was focused on the two NGOs. Interviews were held with project officers in the field and at the headquarters as well as with senior managers and board members. These interviews enabled the clarification of findings as well as the generation of new data. Still, there can be potential problems here too in the form of misleading questions, respondent bias, and reflexity (where the respondent says what the interviewer "wants" to here).

A third source of evidence was that of direct observation. I had the opportunity to observe behavior at both NGOs during 1996 and 1997, at the Manila headquarters and during short visits to project sites in Luzon, Palawan, and the Visayas. While these short visits in particular could not provide the rich insights associated with prolonged residency, it offered nonetheless an additional opportunity to assess "moral capital in action." When suitably qualified in light of possible problems of selectivity (only a few sites visited) and reflexivity (practices possibly distorted by my presence), evidence drawn from direct observation can usefully complement the other evidence.

The case study research presented in this book is an initial exploration of a hitherto neglected subject. It is an attempt to provoke debate and further inquiry into the workings of moral capital. The study of moral capital is thus an attempt to account for at least *some* of the strategic rationality (of a notably instrumental character) associated with the cultural resourcefulness so admired by NGO observers. If it remains to be determined how extensive the impact of moral capital might be in the wider NGO world, the evidence presented here in relation to two Philippine NGOs is nonetheless suggestive. Thus, moral capital is found to be an important force in the determination of political, financial, and territorial strategies of these organizations. By coming to better understand the specific relationship between moral capital and strategic behavior, it might be possible to gain insight into the ways in which cultural resourcefulness can be an important factor in NGO empowerment more generally.

CHAPTER 3

Doing "Good" in the Philippines

However culturally resourceful, NGOs rarely go far in the absence of social misery or environmental degradation. They thrive where things go badly, and hence where moral visions and social prophecies take root most readily. The presence of NGOs in any great number thus bespeaks a land caught between hope and despair, a place of resistance and acquiescence. The Philippines is one such "singular and plural place." It is a place of immense wealth and grinding poverty. It is a global biodiversity "hotspot" and a world leader in environmental degradation. In short, it is a land of contradictions possessing conditions propitious for NGO growth. Not surprisingly, then, the Philippines are renowned for the size and the sophistication of their NGO sector. It is at the forefront internationally in organizing civil society. A land of "broken promises" has spawned an army of the well intended in a curious "dialectic" of good and bad seemingly characteristic of modern Philippine society (Goodno 1989; Clarke 1998; Sidel 1999).

No account of the moral terrain of Philippine life is offered here. Nor is it necessary to provide an exhaustive account of the NGO sector. Instead, I offer a selective outline of the history of Philippine NGOs, drawing on diverse secondary sources and fieldwork insights (e.g., Alegre 1996a; Clarke 1998; Silliman and Noble 1998; Boudreau 2001). The two NGOs featured are reform-minded in orientation and broadly national in ambition if not in territorial reach. Both have enjoyed policy influence even as they have generated political controversy. The two thus have distinctive patterns of development and hence, to reiterate, ought not to be viewed as representative of Philippine NGOs. This is not a problem since, as was noted, the methodology uses a case study approach in which empirical cases are related to theory, not the NGO population (Yin 1994).

Fighting for Change

If NGOs are at the forefront of a global "associational revolution," then Philippine NGOs are at the cutting edge of this process. Whether in terms of organizational diversity, breadth of activity, or sheer numbers, they are often seen as models of appropriate conduct (Broad with Cavanagh 1993; Silliman and Noble 1998). How has this remarkable situation come to pass? A brief overview is in order, and for convenience I divide it into four periods of NGO development.

The first period (1946–1972) witnessed the creation of some of today's largest NGOs, such as the Institute of Social Order (ISO), in 1947, and the Philippine Rural Reconstruction Movement (PRRM), in 1952. The word *reform* summarizes this era, when political and economic leaders sought to modernize the country in the context of a U.S.-led anticommunist struggle. Supported by business, the Catholic Church, and the state, "idealistic professionals began to stream into the countryside to share their technical skills in the fight against rural poverty, while other civic groups mushroomed in Manila" (Alegre 1996c: 7). In keeping with fashionable modernization theory, poverty was viewed as a technical problem amenable to expert resolution. It was not seen as a social condition arising from political and economic inequity. Further, little thought was given to the environmental implications of modernization as forests were felled and lands were degraded with alarming rapidity. Optimism was the order of the day as middle-class professionals assisted the poor. By the late 1960s, however, optimism had faded as the social and ecological costs of development became apparent (Kummer 1992; Broad with Cavanagh 1993; Alegre 1996c).

The second period (1972–1986) coincides with the Marcos dictatorship and was marked by growth in the NGO sector as well as political oppression. Inequalities widened as a result of "crony capitalism" (Wurfel 1988; Boyce 1993; Vitug 1993; Ross 2001). It was a time of deep skepticism for most NGOs as the prospect of equitable development receded. And yet, the dictatorship generated opportunities for NGOs as faith in the ability of the state to act in the public interest declined. Indeed, the NGO sector resembled a surrogate state at times as its employees fought human rights abuses and tackled poverty with the help of clergy, community groups, selected business leaders, and donors (Alegre 1996c). NGOs also confronted the environmental legacy of years of rampant exploitation, with some emphasizing the need to protect residual flora and fauna and others stressing the need to safeguard local livelihoods. This activity took place in a disjointed fashion. However, controversial projects sponsored by Marcos in the 1970s prompted NGOs to form the first national environmental network, the Philippine Federation for Environmental Concerns (PFEC), in 1979 (Ganapin 1997).

Tense relations between NGOs and Marcos overshadowed these activities. Most NGOs sought to distance themselves from a discredited regime for fear of being tainted by association with the despised dictator. Yet, because they were seen as a challenge to the status quo in official circles, NGOs had to tread carefully during this period. The imprisonment of NGO leaders such as Isagani Serrano and Horacio Morales, the torture or murder of activists, and the closure of "subversive" NGOs highlighted the coercive power of the

Marcos state. Great skill was thus required to challenge that state while avoiding the heavy hand of the law—even as that heavy hand was somewhat moderated as a result of pressure from external donors. In the lead-up to Marcos's downfall after the assassination of opposition leader Benigno Aquino in 1983, more and more NGOs joined the anti-Marcos campaign. In the process, they acquired reputations as champions of democracy that stood them in good stead in the post-Marcos era (Goodno 1989; Thompson 1995; Alegre 1996c; Clarke 1998).

The third period (1986–1992) covers the term of President Corazon Aquino and was a time of relative optimism for many NGOs. It started with great expectations for social change, especially with regard to the thorny issue of land reform. Yet prospects for equitable development had dimmed by the end of the 1980s as ruling clans asserted their power in Congress. The Aquino era was nonetheless a time of political influence for reform-minded NGOs. The 1987 Constitution explicitly recognized the development role of the NGO. The Aquino cabinet met routinely with NGOs on social and environmental matters. Indeed, NGO leaders such as Fulgencio Factoran, Celso Roque, and Karina Constantino-David were appointed to top posts in government (Putzel 1992; Gutierrez 1994; Constantino-David 1997).

The alacrity with which Aquino embraced NGOs had much to do with the propensity of donors to provide assistance to these groups. Here was the idea that NGOs were more in touch with local communities than state agencies. One thing was for sure: rapid growth in funding prompted a dizzying expansion in the NGO sector as existing organizations became larger and organizations were created (Constanino-David 1992; Broad with Cavanagh 1993; Alegre 1996a; Clarke 1998). Not all NGOs seemed to be bona fide, however. The creation of "mutant" NGOs was a source of concern, as it was feared that the generally good reputation of many NGOs might be undermined as a result of the actions of mutants widely seen to be fronts for powerful political and economic leaders. That the boundary between "mutant" and "genuine" NGOs could be at times fuzzy only served to intensify that anxiety (Korten 1994; Bryant 2002b). One outcome was the creation of coalitions at the turn of the decade. Coalitions such as the Green Forum, the Philippine Environmental Action Network (PEAN), or the Caucus of Development NGO Networks (CODE-NGO) were created in the belief that the political and ethical interests of NGOs needed defending against mutants as well as other external threats (Aldaba 1992; Legazpi 1994; CODE-NGO 1995).

Although environmental considerations loomed large under Aquino, her government's response was nonetheless ambiguous as it sought to reconcile competing interests. It supported the idea of environmentally sound development and solicited NGO involvement in policy reforms leading up to the Earth Summit held in Rio de Janeiro in June 1992. Yet all seemed to be business as usual in that Aquino proved unable or unwilling to challenge vested interests linked to resource over-exploitation (Ganapin 1989; Vitug 1993; Braganza 1996). This ambiguity left many NGOs unsure of the government's commitment to change. However, the lack of a political alternative—the left was severely divided at this time—meant that many NGOs were at a loss as to what to do other than to intensify political lobbying (Rocamora 1994; Alegre 1996c; Boudreau 1996; Clarke 1998).

This quandary intensified in the fourth period, marked by the term of President Fidel Ramos (1992–1998). Relative political stability and economic growth characterized this period as government sought to industrialize the country by the year 2000. While this strategy earned Ramos praise in pro-market circles, it also led to domestic political tension, as the social and environmental costs of accelerated development became known. Many NGOs were consequently torn over whether to cooperate with Ramos (Nozawa 1996). True, his government was formally committed to achieving sustainable development—hence the 1996 publication of *Philippines Agenda 21* (PCSD 1996). Yet many felt that "action hardly matches the rhetoric" where Ramos was concerned (Serrano 1994: 13).

A common response was critical engagement: a process whereby an NGO works simultaneously with and against state agencies. Thus many pursued official links in the context of community-level projects. From time to time, however, conflict erupted when NGO and state partners attempted to pursue different priorities. Some NGOs challenged overall policies on indigenous peoples or environmental conservation (Kalaw 1997; Rood 1998; Hilhorst 2003). Organizations were also on the lookout for policy reversals. Such was the case when in 1998 the Ramos government lifted the ban on timber exports that had been put in place under Aquino (Dauvergne 1997; Beja 1999). And yet, this fourth period is one in which NGOs tended to deemphasize "big" political issues as they became absorbed in local-level projects. Political action was often of a reactive nature as NGOs sought to counter specific development initiatives believed to be inimical to local livelihoods and environmental sustainability.

From a historical overview ending in mid-1998 with the handover of power from Ramos to Estrada, let me now discuss selected themes in the development of the Philippine NGO sector. A first theme, widely remarked in the literature, is that of organizational "scaling up" (Edwards and Hulme 1992). When funding grew rapidly after 1986, much of this money went to those NGOs with track records and international connections. The rise of NGOs like the PRRM, the Task Force Detainees of the Philippines (TFDP), or the Haribon Foundation, with medium to large work forces and big grants, became a distinctive feature of NGO life (Alegre 1996a; Clarke 1998). Yet critics decried the "swollen" bureaucracies and the "remoteness" of these NGOs from local communities. Their size was believed to make them "less flexible, creative and participatory than smaller organizations" (Constantino-David 1992: 139). This sort of criticism smarted and led these relatively larger NGOs to monitor the effects of their actions on smaller NGOs and people's organizations. They also assessed their organizational structures to look for ways in which to ensure that size did not limit flexibility in the field.

A second theme is the increased number of NGOs operating in the country after 1986. Figures tend to vary depending on the definition of an NGO. Clarke (1998: 69–72), using official data from the Securities and Exchange Commission, suggests that there were 70,200 NGOs as of December 1995—up from 23,800 in January 1984. These figures are substantially inflated, though, by the inclusion of people's organizations in the tally. As such, the number of "genuine" NGOs may be only 10 percent of the total registered number. For our purposes, what is of interest is simply the very rapid rate of growth in NGO numbers. This growth would tend to indicate that the NGO world became crowded in a short space of time.

What are we to make of this change? A growth in NGO numbers resulted in added competition for funds. True, this problem was less acute in the late 1980s and early 1990s when donors put money into the country (Korten 1994; Alegre 1996c). Yet by the mid-1990s some donors were reducing disbursements to the Philippines. The creation of NGO-managed funding bodies such as the Foundation for the Philippine Environment (FPE) and the Foundation for a Sustainable Society (FSSI) provided some reassurance of long-term funding to the sector (CODE-NGO 1997). Overall though, the art of fund-raising became ever more crucial to the welfare of NGOs as the decade unfolded. The increased numbers meant that there was continual pressure to maintain a distinctive identity in a crowded field. Even established organizations were not beyond difficulty, as discussion of the Haribon Foundation will illustrate. This question of identity is closely linked to the problem of reputation—how to earn and capitalize on a good name.

A third theme relates to the issue of professional development. Since donors and others saw NGOs as central to "sustainable development" after 1986, there was a sense that NGOs ought to become more "professional." It was no longer sufficient for them to be seen to promote moral visions. Now, they also needed to pay attention to the *manner* in which they worked. Yet this effort was resisted by those fearful of its corrosive effects on NGO beliefs. There was felt to be a danger in buying into a set of business practices that might be inimical to those guiding NGOs. Instead, the challenge was to "devise [management] systems that are consistent with their very nature and resist the imposition of practices of the establishment" (Constantino-David 1997: 23). To succumb to the latter might also undermine the ability of NGOs to accumulate moral capital with local communities and other partners distrustful of business and government.

A fourth theme concerns the problem of so-called mutant NGOs. When elites created these NGOs to mimic the form and style of regular NGOs, they did so in order to obtain various social and economic benefits. Pseudo-NGOs aim to be seen to pursue moral concerns in "altruistic" fashion—the better to promote highly partisan interests (Bryant 2002b; Top 2003). In the process, though, they began to give the NGO sector as a whole a bad name as "scams" were reported in the media. Regular NGOs responded by creating coalitions to better define their identities and to defend their interests. CODE-NGO was the most notable case in point. A central goal here was to develop a code of conduct to define how members should behave toward one other, local communities, donors, and state agencies, as well as employees. The objective was to "preserve the integrity" of NGOs by establishing ways to "ferret out" mutants, sustain

NGO commitment, and improve the quality of work in the sector (CODE-NGO 1995: 5). True, the coalitions were created for other reasons. They were formed to strengthen the position of NGOs in relation to state agencies as well as to enhance the social profile of NGOs. They also served to build cooperation among like-minded groups. Yet the period of most frenetic coalition building coincided with the greatest proliferation of mutant NGOs, thereby emphasizing insecurities about identity and the need for "common action" (Constantino-David 1992: 144).

The discussion thus far has suggested that the development of the Philippine NGO sector is partly conditioned by the actions and perceptions of other actors. If we think back to the discussion in the first two chapters, it is possible to begin to see how the interplay of perception and power, reputation and strategy plays out on the grand canvas of Philippine NGO politics. How, though, does this general discussion relate to the practices of individual NGOs? What are the implications for specific organizations of thinking about NGOs as moral entrepreneurs in pursuit of moral capital?

Sociopolitical Aims, Ecological Outcomes

The two NGOs featured in this study illustrate, among other things, how the pursuit of social justice and environmental conservation has tended to converge over time in the Philippine NGO sector. It is the rare development NGO that does not now acknowledge the environmental underpinnings of social justice, while it is the rare environmental NGO that does not similarly today recognize social and political preconditions for environmental sustainability. If anything, this process has reinforced the ambiguity of NGO action in the modern world. The consolidation of uneven global capitalism under neoliberal conditions has multiple if contradictory impacts on NGOs. The entwined sociopolitical and environmental practices that many promote are increasingly affected by market dynamics (Sogge 1996a; Fowler 2000). Yet the ability of NGOs to act becomes ever more dependent on a series of compromises with pro-market groups such as business foundations, donors, state agencies, and even international NGOs. All of which is to say that many NGOs have strong incentives to be moral entrepreneurs as they weigh the consequences for reputation of moral dilemmas that condition their existence, as we will see. Let me now introduce the two NGOs featured in this analysis.

Philippine Association for Intercultural Development (PAFID)

The PAFID assists indigenous peoples "in their quest to regain and protect their ancestral domains and struggle for self-determination" (PAFID 1994c: 1). Although social concerns are privileged, there is recognition that the search for social justice and environmental welfare are linked. Consider the NGO's vision as articulated in 1994: "The PAFID views Tribal Filipino communities as responsible stewards of their resources. As communities and individuals, they possess land rights and responsibilities, cultural integrity, educational and health systems that embody their indigenous knowledge and mores, and employ socially and ecologically sound methods of managing and utilizing their natural resources to their own and national benefit for both present and future generations" (PAFID 1994c: 1).

The organization has earned a national reputation based on practices related to this moral vision. Yet the process has not been easy, due to the vicissitudes of institutional life as well as challenges arising from the wider political economy. The PAFID was registered with the Securities and Exchange Commission in Manila on August 1, 1967, as an NGO dedicated to assisting "tribal" communities. The brainchild of a group of scholars, lawyers, and indigenous leaders, it started out with a seven-person board, one full-time employee, and a core aim to help communities threatened by ill-conceived development. But no indigenous group sought help. Accordingly, and following the resignation of director Nery Diaz-Pascual in 1970, the PAFID effectively ceased to exist. As the PAFID's Delbert Rice (1996) recalled, the NGO's "vision was bigger than its facility" for action at this time.

The organization received a new lease on life in 1975. Concerned about the plight of communities under Marcos, a different group met to see what could be done. They considered founding a new organization, but one of them remembered the "sleeping PAFID." The original board was contacted and agreed to revive the NGO. A new board was elected, initial funds were obtained, and a staff was appointed. The organization immediately launched itself into the political struggles racking northern Luzon, a region home to many of its employees and targeted for "mega-projects" such as the Cellophil logging-cum-industrial processing project and the Chico Dam (Anti-Slavery Society 1983; Hilhorst 1997, 2003; Top 2003).

These struggles led to a deep ideological split within the NGO. While a leftist faction advocated political confrontation, a politically centrist faction pushed a program of "practical" and "nonpolitical" action to assist indigenous partners while avoiding a military crackdown. The split weakened the NGO just as it intervened in the Cellophil controversy. In a highly charged atmosphere, at least one employee defected to the Communist-linked New People's Army (NPA). The political moderates—among whom was board member Delbert Rice—were left in control, but they thought it prudent to "put the NGO back to sleep" until more propitious times (Rice 1996).

This second hibernation was short-lived, in that the organization was revived in the early 1980s. University graduate Sammy Balinhawang got things moving again with the help of Rice. Two additional graduates were hired to sort out the PAFID's mission, field operations, and funding. Great caution was needed, though, at a time of political and economic crisis. Official links were selectively cultivated so that the NGO could maintain its campaign on behalf of indigenous peoples. Land tenure was the key issue. The PAFID thus sought Communal Forest Leases (CFLs) for communities through the Department of Agriculture and Natural Resources (forerunner of the Department of Environment and Natural Resources or DENR). Donors funded this pioneering work in order to promote reform. These included the Ford Foundation, Philippine Business for Social Progress (PBSP), and the U.S. Agency for International Development (USAID). Yet the PAFID also sought to avoid being too closely associated with donors such as USAID for fear of NPA retaliation. The nature of this NGO's work—policy advocacy and fund-raising in Manila, combined with fieldwork in remote areas subject to NPA control—meant that it had to tread a fine line in dealing with pro- and anti-Marcos forces.

The 1986 revolution prompted great change for the organization. New democratic freedoms and enhanced funding meant that it was better able to pursue its mission. The PAFID's pioneering status translated into a high level of credibility with donors, state officials, and local communities alike. At about this time too, the NGO incorporated environmental concerns explicitly into its activities. This change reflected the recognition that land tenure alone was not sufficient for the development of sustainable indigenous livelihoods. It was also, to be sure, a canny move to widen donor appeal at a time when sustainable development was popular.

Institutional development was thereafter rapid. Funding was obtained from such donors as the Philippine-Australian Community Assistance Program (PACAP), Misereor, the Canadian Hunger Foundation, and the Asian Development Bank (ADB). Income climbed from a few hundred thousand pesos per annum to several million pesos per year. There was also dramatic growth in staff numbers. Whereas there were but three full-time employees in 1985, by 1993 that number had leaped to nearly fifty. A sizable proportion of this increase was accounted for by one large ADB-funded project (see chapter 5). Still, these hefty increases enabled the NGO to develop new territorial bases as well. Thus work continued in traditional areas of strength in Luzon and Mindoro even as projects were initiated in Palawan, the Visayas, and southern Mindanao. In the process, the PAFID became one of the leading NGOs working on indigenous issues in the country.

Relations between the PAFID and selected state agencies grew closer. The NGO redoubled efforts to secure indigenous land claims at a time when indigenous concerns received unprecedented political attention. By September 1992, it had helped 7,866 families to secure land covering 66,525 hectares—most of this assistance occurring after 1985 and through DENR Community Forest Stewardship Agreements (CFSAs). Cooperation was also reflected through donor-funded projects involving the DENR, the PAFID, and people's organizations working together on land tenure and other problems. Finally, there was the practice of having a senior DENR official on the NGO board (PAFID 1993b; see also Rood 1998; Bryant 2000; Gauld 2000).

The mid-to-late 1990s were a period in which the PAFID's operations were consolidated. The goal was to concentrate energy on land tenure and downplay other activities such as livelihood support. Relations with DENR officials were somewhat tense as a result of the pullout from a major project in Mindoro in 1993 (see chapter 5). This event also meant that the organization had to lay off staff, causing some anger in the process. Those who remained underwent training on policy and legal development, agro-forestry, geographical information systems (GIS), or accounting. The "professionalization" of the PAFID mirrored broader trends in the NGO sector (Fowler 2000).

There were two further trends in institutional development in the 1990s. A program of decentralization sought to boost local-level decision-making and responsiveness. Three regional offices were thus established, with each enjoying considerable operational autonomy. The change reflected the aforementioned concern that the size and broad purpose of "larger" organizations might hinder their effectiveness on the ground. It was also a move designed to realign the PAFID so as to take full advantage of state decentralization initiated with the Local Government Code of 1992. There was, too, an emphasis on policy advocacy. PAFID's Dave De Vera and other Manila-based staff were active in the campaign to pressure Congress to recognize ancestral domain. There was thus the promotion of the Mangyan Ancestral Domain Bill (or "Mindoro Bill"). Further, nationwide ancestral domain legislation was also sought. A victory of sorts was achieved when the Indigenous Peoples' Rights Act (IPRA) was passed in October 1997 along with Implementing Rules and Regulations in the following June (Congress of the Philippines 1997; National Commission on Indigenous Peoples 1998).

And yet, strong opposition has persisted to the recognition of indigenous rights. Indeed, progress in turning the new law into rights was limited. At the heart of this reluctance to move forward on the IPRA was the emphasis of successive Philippine governments on large-scale mining as a means to accelerate development. This view was reinforced as the Asian financial crisis of the late 1990s jeopardized modest gains made under Ramos. The basis for action here has been the 1995 Mining Act, which is widely seen as a model piece of neoliberal legislation solicitous of foreign capital (Nettleton 1996a, 1996b). Since key mineral sites are on land claimed by indigenous people, the struggle over ancestral domain intensified. The PAFID has persisted with political advocacy on behalf of indigenous people (Environmental Science for Social Change 1999).

On the occasion of the NGO's twenty-fifth anniversary in 1993, it published a study on communal land titles for indigenous Filipinos. The book neatly summarized the thrust of the PAFID vision and argued that such titles would "meet the land tenure problem" confronting indigenous people even as it would "benefit the nation as a whole by encouraging the tribal people to act as effective stewards of their resources which are, at the same time, national resources" (PAFID 1993a: i). The appeal of the PAFID has thus derived from a focus on one key issue—land tenure—to argue a wider case for social and environmental justice.

Haribon Foundation

The Haribon Foundation, in contrast, has derived appeal from its role as a pioneer among environmental NGOs. From humble origins in the early 1970s, it became one of the largest and best-known NGOs in the country, renowned abroad too for its environmental research and political lobbying. In its heyday during the anti-logging struggle of the late 1980s and early 1990s, Haribon was at the forefront of a broad coalition of social groups dedicated to ending a political economy of natural resource depletion.

Political activism was far from the minds of the founders. Established as a bird-watching society in 1972, the organization sought to protect the endangered Philippine eagle after which it was named. The focus shifted in the late 1970s to encompass wider environmental concerns. If this change raised questions about the environmental practices of Marcos, political activism was deemed too dangerous at the time. Instead, Haribon used scientific research and discreet lobbying of senior officials to push a conservation agenda. Research was central to this process—not the least because it led to links to such international NGOs as the World Conservation Union (IUCN), the World Wildlife Fund (WWF), and the International Council for Bird Preservation (later Birdlife International).

The "People's Power" revolution prompted a rethinking at Haribon that led to an altered mission. The old strategy—produce scientific data on environmental problems in the hope that remedial state action would follow—had failed. The new strategy linked scientific work to social activism in the quest for alternative development. Thus, a "people's agenda" sought to link destruction of forests and coral reefs to people's livelihoods so as to build a constituency for conservation. It was a good time to revise the mission. A newly elected Aquino sought allies in the NGO sector. A new generation of activists and researchers led by Maximo "Junie" Kalaw joined Haribon in a process that reinforced social aspects to the NGO's work. And international momentum was gathering behind the call for "sustainable development" following the 1987 report of the World Commission on Environment and Development (or Brundtland report as it was commonly known). This particular concept provided an ideal means by which Haribon could link environment and development (Kalaw 1997).

The new activism brought Haribon into conflict with powerful opponents in politics and business. One campaign targeted *muro-ami*, a Japanese fishing method used by unscrupulous companies that destroyed coral reefs and relied on child labor. Such labor turned out to be supplied by a company owned by a senior member of Congress. Thus this campaign provoked a backlash. But this was nothing compared to what would happen when the organization launched the anti-logging campaign. The early focus of the campaign was on rapid deforestation in the biodiversity-rich province of Palawan. Haribon launched a high-profile national signature drive designed to prod Congress into banning commercial logging there. A local membership chapter called Haribon Palawan was founded in May 1989. Yet it was a tough battle. Pagdanan Timber Products was axing a large proportion of the forest. This company, owned by the powerful local boss Jose Alvarez, could count on the support of many politicians and military officers. A campaign of harassment got under way that culminated in the February 1991 arrest of fourteen members (including Haribon Palawan leader Joselito Alisuag) by the Philippines National Police on trumped up charges (Broad with Cavanagh 1993; Arquiza 1996).

The intimidation backfired as the campaign won national and international media coverage. Haribon thus upped the ante by helping to create an NGO coalition, the Task Force Total Commercial Log Ban, in 1990. The campaign for a national ban on commercial logging was in the end narrowly defeated in Congress in the run up to the 1992 general election, but regional efforts to end logging thereafter enjoyed some success. Indeed, in Palawan there was a palpable "environmental turn" as a new governor was elected with NGO support. Although Haribon was not the only NGO that was involved in these battles, it was invariably at the forefront and became virtually a household name among middle-class urban residents in the Philippines.

The new activism was reflected in organizational change. Scientific investigation remained important even as there was a new emphasis on applied research. There was also a focus on community organizing reflecting a belief that local communities were appropriate environmental managers. A significant legal branch was added in 1988 when the *Tanggol-Kalikasan,* or environmental defense program, was created to pursue environmental litigation, legal research, and popular environmental legal education. This program was significant insofar as it revealed an organization keen to take the struggle for social and ecological justice into the realm of law.

The new-look Haribon found relations with the DENR strained. The anti-logging campaign generated some support but also much opposition at a DENR itself divided over policy issues. The high public profile of the Haribon-led campaign added to the tensions. Thus criticism of USAID's Philippines Forestry Sector program by

Junie Kalaw before the U.S. House of Representatives in September 1990 angered DENR Secretary Factoran, who, despite pro-NGO sympathies, nonetheless resented such "interference." Factoran's ire intensified when Kalaw was quoted in the national newspapers as calling for the secretary's resignation in the wake of the Ormoc tragedy of November 1991. In this town in Leyte at least five thousand people had died following flash floods that were linked to deforestation (Severino 1993; Teehankee 1993; Vitug 1993; Bankoff 2003).

And yet, Haribon sought to cooperate in areas of mutual interest with the DENR through a strategy of critical engagement. It was thus the lead Philippine NGO in a debt-for-nature swap program that also involved the WWF and the DENR. The organization also signed up to be part of an eighteen-member NGO consortium as part of the World Bank sponsored and DENR coordinated NGOs for Integrated Protected Areas (NIPA) program. This sort of work was certainly affected by the tensions just noted. Indeed, political fall-out from the anti-logging campaign after 1992 included efforts to squeeze Haribon out of the final phase of the debt-for-nature swap program. Nonetheless, links to the DENR survived these turbulent times.

The Ramos years marked further change at Haribon. As with the PAFID, there was a push to emphasize work at the community level. This shift partly reflected "large NGO syndrome" or the fear of being seen to be "out of touch" with local partners. It reflected, too, disenchantment with national politics following the defeat of the log ban bill in the House of Representatives as well as the poor showing of green candidates in the 1992 elections. Passage of the Local Government Code in 1992 also signaled the growing importance of local governance in environmental matters. Such decentralization only reinforced the feeling at Haribon that local action and support would be essential to future success. Still, if the NGO emphasized community organizing and legal support after 1992, national battles over a proposed cement plant at Bolinao in 1996 and a DENR reversal on the log export ban in 1998 revealed an organization not averse to confronting powerful political and bureaucratic figures when necessary (Beja 1999).

These changes at Haribon were connected in complex ways to shifting financial circumstances. The phase-out of the debt-for-nature swap program happened at a time when a number of other programs were ending. As a result, staff numbers were cut substantially, from eighty in the early 1990s to just fifty by mid-decade. There was also the scheduled phase-out of a major MacArthur Foundation grant, reflecting that donor's reallocation of funds from the Philippine country program to other areas. Since this grant had largely underwritten administrative costs not covered by project grants, the NGO needed to move to a project-based organizational model in order to ensure that expenditure more closely reflected project income. The result was a reduction in headquarters staff in favor of a slight increase in field staff. Executive director Ed Tongson also sought to boost income through a diversified fund-raising strategy that included the sale of Haribon products such as backpacks and t-shirts as well as via a membership drive.

These developments prompted much soul searching at Haribon. True, the NGO had national and international name "projection" based notably on a high public profile during the anti-logging campaign. Yet this was no ground for complacency. Ironically, Haribon's success in drawing attention to environmental issues contributed to its own difficulties in the 1990s. Not only did Haribon's work pave the way for the creation of new generation of environmental NGOs, but it also prompted development NGOs such as the PAFID to integrate environmental concerns into their work. The fact that virtually "every NGO was an environmentalist now" meant that Haribon was under pressure as never before to define a niche for itself in an increasingly competitive context. Still, its strength in scientific research and multiscale policy advocacy, as well as its track record, was some reassurance as this NGO faced an uncertain future at the turn of millennium.

Devising a Game Plan

Nongovernmental organizations such as the PAFID and the Haribon Foundation not only contribute to but also reflect social change. Circumstances can change NGOs just as NGOs can change circumstances. It is this situation that is of interest in the present book. It is a situation that I argue may be partly described by thinking of NGOs as moral entrepreneurs involved in a quest for moral capital. Whether helping indigenous people or fighting for endangered flora and fauna, NGOs such as the PAFID or Haribon are often perceived as promoting a moral vision that benefits others. Yet this is much more than simply a story of good intentions. To help others, NGOs need first of all to help themselves by acquiring the *means* to effect change. Those means include such things as labor and money. But the acquisition of these resources is, I suggest, predicated on a process in which moral capital is accumulated. However, that process is not straightforward, since the uncertainties of social life combine with the vagaries of reputation building to produce a context ripe with ambiguity.

Having a game plan—a sense of how to proceed—is vital under these circumstances. Such a plan is also a useful means by which to assess strategic rationality that may be associated with NGO behavior. Here, political, financial, and territorial calculations are seen as being central to understanding how NGOs strive to convert their visions into reality. To appreciate the nature and significance of NGO behavior is simultaneously to understand why NGOs pursue the strategies that they do. To be sure, the relationship between motives and behavior is complex. However, the careful scrutiny of words and deeds can pay rich dividends in terms of a clearer grasp of what NGOs "stand for" as well as how they might seek to empower themselves through means of a reputation. This is not to suggest that NGOs are aloof from "nonmoral" concerns or even that they are averse to moral compromise. To the contrary, in suggesting that NGOs may orient their behavior around a quest for moral capital, I hope to show how this actor habitually and *simultaneously* negotiates a course of action that is shaped by moral vision and practical necessity as well as self- and other-regarding behavior. The ambiguous nature of the instrumental form of rationality displayed by NGOs in their roles as moral entrepreneurs will become clearer as we consider the interlinked questions of political, financial, and territorial strategy.

CHAPTER 4

Political Virtuosity

A quest for moral capital may be reflected in political strategizing by nongovernmental organizations (NGOs). Political tools are needed to help organizations get the most from relationships with state agencies, community groups, and others. These relationships vary, but general patterns can be identified at least insofar as our two case studies are concerned. They probably apply to a greater or lesser extent to other organizations as well.

A multifaceted approach to the state is developed through critical engagement *and* constructive engagement. The balancing of cooperation and criticism is termed *critical engagement* to emphasize the ambiguous nature of this relationship. Such ambiguity is due in part to the internal complexity of the state, and hence the need to tailor approaches to different agencies and personnel. Yet critical engagement involves more than fine-tuning. It is a potentially efficacious way in which to pursue moral capital with state *and* nonstate partners. Meanwhile, state agents themselves enter into critical engagement with NGOs to further their own ends—notably to harness these organizations into official projects and procedures as well as to boost their own legitimacy (Alagappa 1995; Bryant 2001). Meanwhile, *constructive engagement* describes relations between NGOs and local communities often indispensable to NGO activities. Here, moral beliefs and political calculations mix. Organizations simultaneously seek to implement visions, develop mutually beneficial community links, *and* ensure that they are seen to be fighting the "good cause." While these relations may reflect "win-win" scenarios, conflict can still bedevil this partnership.

These two forms of engagement reflect *predominant* tendencies in the core political relationships under review in this chapter. They do not capture all facets of those relations, let alone other relationships in which NGOs become involved. Yet the sorts of engagement described below are useful shorthand for assessing more generally how political interaction may influence a quest for moral capital by our case study NGOs.

Seizing Political Opportunities

The ability to devise strategy is predicated on certain political preconditions. The fate of NGO plans is conditioned in part by shifts in political environment—what political sociologists call political opportunity structures. Tarrow (1994: 85) suggests these are "consistent—but not necessarily formal or permanent—dimensions of the political environment that provide incentives for people to undertake collective action by affecting their expectations for success or failure" (see also Kitschelt 1986; McAdam et al. 1996). Our concern is with shifts in resources *external* to an NGO that affect its prospects—things like altered political alignments, new allies, divisions among opponents, and wider access to decision-making structures. Through relating this notion to political change in the developing world, it can be seen how there are limited prospects for action under strongly authoritarian regimes, but greater prospects under more democratic ones. NGOs, in this view, can best work when there is freedom of association and speech.

That political opportunity structures condition NGO prospects suggests that these organizations cannot defy gravity. Yet it is wrong to ascribe too much significance to them. NGOs can thrive in inhospitable settings—as we have seen (see also Eldridge 1995; Clarke 1998; Bryant 2001); further, the structures are not rigid. As Jasper (1997: 36) notes, "the term *structure* misleadingly implies relatively fixed entities, so that attention is often diverted away from open-ended strategic interplay." Indeed, NGOs have helped to shift these structures, not least in the Philippines where they are potential "vehicles of democratization" (Silliman and Noble 1998b: 7; see also Heyzer et al. 1995; Fisher 1998).

It is important to note two things here. First, while political opportunity structures rarely determine the details of NGO life, they do condition the networks through which moral capital is accumulated. For example, in the Marcos era reputation building was mainly a question of working with community organizations and donors, inasmuch as the state was held in disrepute. Some feared that moral stigma would attach to any NGO seen to be too close to the dictator. As one veteran observed, NGOs sought "moral ascendancy" through speaking for those unable to speak (Francisco-Tolentino 1997). The nature of political opportunity structures will thus condition not only which political opportunities exist but also which opportunities NGOs choose to pursue. Second, these structures may condition the context in which NGOs operate even when they are seemingly freest. Since 1986, they have experienced few constraints when compared with conditions under Marcos. True, democratic circumstances remain contingent, yet there is undoubtedly greater room for maneuvering. This offers a greater array of possibilities for NGOs in strategizing over moral capital. With whom should they associate and under what terms? How should NGOs behave toward a state committed to formal democratization?

Political strategizing is at its most interesting here precisely because choice is greater, and hence, the need to make difficult tradeoffs.

That this study uses evidence drawn largely from the period 1986 to 1997 shapes its findings, given that political opportunity structures were characterized by quasi-democratic conditions then. I am thus not concerned with understanding how dramatic shifts in structures might affect the circulation of moral capital (Jasper 1997; Della Porta and Diani 1999). Instead, focus is on a period characterized by flourishing moral capital networks to understand how NGOs devise political strategies with an eye to accumulation. That strategizing may look different under altered circumstances is not to be gainsaid and is for future research.

The Problem of the State

Many NGOs are suspicious of political and bureaucratic leaders even when democratic conditions exist. Such suspicion is partly a legacy of practices perpetrated in intolerant times, especially if perpetrators go unpunished. There is also the fear that a corrupt official or politician might undermine fragile freedoms (Eccleston and Potter 1996). Other factors relate to the development of the NGO sector. That development is complex but is linked to perceived state failure. Much of the history of the NGO sector is about undertaking work that states do not do (well): health care and livelihood support for the poor, or environmental conservation, for example. Since the fall of Marcos, the state has been subject to new pressure (not least due to NGO campaigning) and has sought to address some commitments. As a result, NGO strategizing needed to be more complex, with critical engagement the order of the day. Born of an ambiguous NGO-state relationship, it encompasses both cooperation and conflict over social and ecological concerns. The regulation of conflict is crucial and needs ring-fencing to avoid having conflict spread—hence the apparent prevalence of targeted criticism of specific practices.

When political opportunity structures acquire a democratic hue, the response by political and bureaucratic leaders to NGO criticism tends to be less hostile. The acceptance of criticism may bolster official legitimacy inasmuch as it "demonstrates" political maturity (Alagappa 1995). Criticism may also be worth enduring if cooperation with the NGO sector enables agencies to tap expertise and funding (Ross 1996). As NGOs work with the state, leaders can anticipate less political opposition as energy is channeled through official practices (Bryant 2001). How might NGOs benefit from critical engagement? I consider below the experiences of the PAFID and the Haribon Foundation to address this question. The picture emerging from the discussion is one of behavior characterized by the building of routine interaction on the one hand and the maintenance of critical distance on the other.

Building Routine Interaction

A key development after Marcos was the elaboration of routine interaction between NGOs and state agencies. There was engagement through projects, seminars, correspondence, staff exchanges, and so on. These "everyday forms of bureaucratic exchange" seem to have become an element in the quest for moral capital. The search for stability was an outcome of the climate of fear and hostility that typified relations under Marcos. True, some NGOs maintained official links then. The Philippine Rural Reconstruction Movement (PRRM), Philippine Business for Social Progress (PBSP), and the Haribon Foundation, for example, worked to some extent with agencies to promote agrarian reform, rural well-being, and wildlife protection. Yet life even for them was difficult. Negligible political influence meant that NGOs were forced into being reactive as official projects of various kinds threatened communities. In the 1970s, for example, Marcos targeted the mountainous Cordilleras in Luzon. Hydroelectric dams were proposed for the Agno and Chico rivers, while logging by the Cellophil Corporation devastated the lands of the Bontocs, Tinggians, Ibalois, and Kalingas. Organizations like the PAFID "campaigned hard" in these areas, using official contacts to promote peace. However, as the PAFID's Delbert Rice recalled, it "lost this battle," and violence involving the army, indigenous people, and the communist New People's Army (NPA) intensified (Rice 1996; see also Anti-Slavery Society 1983; Hilhorst 2003).

There was also a price to be paid for crossing Marcos. Opposition to the sort of projects just noted was met by official harassment, including surveillance, arrest, and murder. This daily experience hardened anti-Marcos feelings. It also required that activists be low-key. At the PAFID, connections to the Roman Catholic and Protestant hierarchies were used to hold meetings on church premises. Minutes of meetings were rarely kept for fear of their falling into government hands. Still, Rice and some other colleagues felt that the organization was too embroiled in political controversy for its own good, such that it was "put back to sleep" in the late 1970s to await better times (Rice 1996). And yet, NGOs such as the PAFID were spared even worse depredations due to Marcos's need to curry favor with international donors (especially the United States) and the pressures emanating from these quarters to "go easy" on the NGO sector (Wurfel 1988).

Thus routine interaction became possible only after President Aquino was in office. I noted how the appointment of NGO leaders to government posts and reference to the NGO sector in the 1987 Constitution signaled that change was afoot. However, it was mundane links between NGOs and state officials that advanced relations. Policy differences and bureaucratic intimidation persisted. Yet post- 1986 political change was a basis for cooperation with reform-minded NGOs at least. Increased donor funding greased the wheels of routine interaction as funds provided a focus for cooperation and justification for closer links (Ross 1996). At the same time, these were new opportunities to accumulate moral capital with state and non-state groups alike.

Transforming agencies into "NGO-friendly" spaces was a pre-condition for routine interaction. For instance, the DENR created dedicated NGO desks to facilitate involvement in its activities. Their creation also signaled a commitment to putting past bureaucratic practices behind them. In the process, the DENR created rules governing NGO involvement through an accreditation scheme. That scheme was to "select NGOs which can best address the most pressing needs and problems of the masses" in a responsive manner so as to help the masses "increase their confidence and self-reliance" (DENR 1992 art. 3, sec. 9). It reflected in its formal language and aims at least an attempt to incorporate democratic ideals into practice. It also defined what a "good" NGO was. Thus NGOs with "integrity and commitment" and that were "reputable and socially acceptable to the concerned and/or affected community" were needed—provided that they were "locally based and with adequate basic resources," "technical capability," and a "proven track record" (DENR 1992 art. 3, sec. 9).

Officials could use this checklist to verify reputations and weed out disreputable NGOs such as those behind the contract reforestation fiasco of the early 1990s (Korten 1994; Bryant 2002b; see also DENR 1994). More positively, the list could help to select the "best" NGOs to work with. The criteria were suggestive of a sort of thinking at the DENR as to the perceived advantages of NGOs: proximity to communities, organizational capacity, and social dedication. At the same time, the criteria seemed to indicate upon what basis the DENR might create moral capital for NGOs. Clearly, these criteria do not tell the whole story and do not encompass a wider set of calculations involved in DENR-NGO relations (Clarke 1998). Nevertheless, routine interaction as described has been important—and has been partly about the agency specifying in formal terms what it values about, and expects from, NGOs.

Such cooperation links NGOs and the DENR in the effort to halt environmental degradation. This has usually involved an arrangement whereby the DENR joins with an NGO and people's organization to promote community resource management. Such cooperation has not been without difficulty (Bryant 2000; Gauld 2000; Lawrence 2002), and the experience of the Haribon Foundation in the Mount Isarog National Park project is illustrative. Mount Isarog is a site of biological diversity, a regional source of water, and home to numerous poor farmers. Over time, the human impact has intensified to such a degree that logging, agricultural encroachment, and habitation have jeopardized the very properties that led to the area's designation as a national park in 1938.

Haribon first became involved there in 1989, when the student chapter conducted a biological survey. The NGO played a key role in community organizing, paralegal training, and promoting alternative livelihoods. What is of interest now is that this work placed it in contact with the DENR Protected Areas and Wildlife Bureau (PAWB), whose job it was to protect the park. Three things stand out here. First, routine interaction involved regular meetings to share information on such things as illegal logging or community organizing. One such forum was the Protected Areas Management Board (PAMB). There were also meetings at DENR headquarters when problems emerged.

Second, much interaction involved Haribon's prodding the DENR to act against illegal loggers identified by the NGO and people's organizations. Criticism was indeed a staple of these relations in the early 1990s as activists charged the DENR with failing to arrest loggers. One Haribon officer noted that relations grew strained when the NGO accused one official of complicity in the logging (Resurrecion 1996). The DENR denied these charges, arguing that "logistically, it was impossible for them to make the rounds and arrest the illegal loggers" due to a lack of staff and funds (Rivero 1996).

Finally, and the war of words notwithstanding, routine interaction was a useful basis by which Haribon could build its reputation with local officials. Many of them supported the broad thrust of Haribon's work since it would reinforce their own bureaucratic interests even as support for a reform-minded NGO might take the sting out of the NPA insurgency in the area. For these reasons, some were happy to praise Haribon as "a credible NGO that deserves the credit for the decline in illegal logging activities in Mount Isarog" (Araojo 1996; see also Nuyda 1997). True, the reports were not all glowing, and the NGO earned the enmity of barangay (or barrio) captains involved in illegal logging (see below). And yet, the bumpy road to cooperation here seemed to pay dividends in terms of enhanced moral capital at the local DENR.

Routine interaction is often involved when NGOs seek to change natural resource regulations. The DENR Certificate of Ancestral Domain Claims (CADC) program, the focus of much NGO attention in the 1990s, is illustrative. This program formally acknowledged the wish of indigenous people for resource control, enjoining them to develop ancestral management plans premised on "cultural integrity" and "traditional resource rights" (DENR 1996: sect. 3). While not abolishing existing resource agreements, it was nonetheless an opportunity for local input into decision-making as a first step toward full ancestral rights (Utting 2000; Bryant 2002a).

The CADC was a means through which the PAFID and others could assist indigenous peoples. This work involved close consultation with communities). It also meant, as PAFID leader Dave De-Vera (1996) explained, a need "to build up skills" in dealing with the

DENR. Those skills include understanding legal and bureaucratic dynamics so as to be a policy advocate. The latter has ranged from discreet lobbying of officials, through formal correspondence supporting claims, to public rallies designed to prod the DENR to act. As the PAFID staff became experts on CADCs, they were drawn deeper into close interaction with DENR counterparts at the local, regional, and national levels. As one senior official remarked, the result was that the DENR was "forced to work with NGOs" just as NGOs were "forced, in return, to work with the DENR" (Austria 1997; see also Leonen 2000).

Even as NGO-state relations are normalized through common endeavor, conflict still occurs as NGOs pursue moral capital. How might an NGO attempt to ensure that conflict does not escalate? Much depends, as we shall see, on how criticism is presented. Routine interaction itself can be a safeguard here—that is, the meetings, workshops, letters, and reports that constitute everyday forms of bureaucratic exchange. While this communication is vital to the production of outcomes, it can also solidify relationships as trust and mutual understanding develop. The accumulation of moral capital is thereby facilitated. In the process, it becomes likelier that state counterparts will not take to heart NGO criticism.

Interpersonal solidarity is central here. Routine interaction is bolstered through the development of personal connections (or "social capital"). One manifestation of this process is when officials serve on NGO boards that help to govern an organization. As Tandon (1995: 42) observes, a board focuses on "issues of policy and identity" to ensure the NGO's "effective functioning and performance in society." At the PAFID, board members are active in providing advice and support on hiring, grants, financial accountability, and working with the DENR. On the latter, the DENR's Joey Austria was invaluable—and illustrates how moral capital is gained through personal links. Austria joined the board soon after the election of

Aquino, when the DENR was being prodded into improving links to indigenous people (Rood 1998; Utting 2000). As a senior advisor on indigenous people at the agency, he led in promoting indigenous land tenure there. He sought to build a role for the PAFID in this endeavor, thereby enabling it to pursue a mission effectively. There were various aspects here. Austria arranged for PAFID staff to be deputized by the agency in order to hasten the processing of land claims. He also kept them briefed on new initiatives—for example, on the new CADC system in 1994. Finally, Austria intervened on occasion to soften the impact of PAFID criticism of DENR practices. Thus, in one confrontation in Mindoro in the early 1990s, he ensured that correspondence was couched in a way that would minimize offense (PAFID board minutes 1992–96).

Such personalized connections sometimes led to problems. For one thing, NGO actions may embarrass officials on the board. Austria was once discomfited, for example, when a PAFID employee in Davao accused DENR counterparts of corruption. It fell to him to smooth ruffled feathers by stressing that the accusation did not reflect the position of the PAFID as a whole (Austria 1997). For another thing, success at linking an NGO to the DENR can provoke opposition within an agency divided into factions. Austria's interventions provoked a campaign by opponents wed to a pro-logging position. He was accused of "receiving money from PAFID" because the NGO paid travel expenses (Rice 1996).

Personal bridge building can also be seen when individuals move between sectors through career development. As the Haribon Foundation found, though, this can be something of a double-edged sword. The NGO staff had hoped that influence in government would increase when their former president, Celso Roque, became DENR undersecretary in 1987. Yet this did not happen because of personal antagonism between Roque and the new Haribon president, Junie Kalaw. Roque disliked the NGO's role in the anti-logging campaign spearheaded by Kalaw. In Roque's view, Kalaw's attacks on the DENR were ill judged since the agency could not "cancel logging concessions overnight" (Roque 1997). In the end, he resigned from the board, in part due to such criticism and in part because he was "not being told what was going on" at the NGO (Roque 1997). A different dilemma occurred when President Ramos appointed Haribon board member Angel Alcala to the top post at the DENR upon the urging of NGOs. However, elation at Haribon soon turned to despair. Rumors that Alcala's associates were corrupt led to media investigations (Severino 1995). As the scandal deepened, the Haribon leadership became alarmed that their good name was being tarnished. Thus, when Alcala was forced to resign in July 1995, Kalaw thereupon impressed on the new DENR secretary Haribon's disapproval of Alcala (Haribon Foundation board minutes 1995–96). Indeed, he even called for a special commission to investigate the affair in a move later dismissed by Alcala (1997) as "opportunistic."

In assessing the significance of routine interaction in structuring NGO-state relations, it needs to be recognized that such interaction represents a *strategic choice* on the part of NGOs (and, for that matter, state actors). If routine interaction is one way to pursue moral capital, though, how might it aid this process and what benefits might ensue? Central here is its utility in boosting capital with state partners. True, reputations are buffeted by conflict between factions in government inasmuch as NGOs become identified with specific individuals. That said, the ability to generate a track record provides organizations with a basis for long-term cooperation above and beyond the influence of mentors. For both Haribon and PAFID, a high standing has been achieved notably due to their pioneering efforts, extensive connections, and effective work. For Haribon, a track record for scientific community organizing and paralegal work has ensured that it has been able to "ride out" periodic tempests. For the PAFID, being a key proponent of land tenure for indigenous people has facilitated their standing.

To what end, though, does this favorable situation lead? The main benefit would seem to be policy influence. Yet there are definite limits to influence, due to fragmented policy-making and countervailing influence by opponents. The PAFID experience with assisting the Tagbanua of Coron Island (Palawan) is illuminating in this regard. The aim was to help them acquire a CADC so as to strengthen efforts to stop unwanted economic activity in the area. PAFID involvement with the community and its Tagbanua Foundation of Coron Island (TFCI) went back to the mid-1980s. The NGO sought to use a good name in government to halt development intrusions in favor of the ancestral domain option for Coron Island. This was always going to be an uphill battle. A local DENR official supported illegal logging while the mayor supported mass tourism for the area in a plan backed by the Department of Tourism. And, a conservation program backed by the European Union and the DENR advocated an integrated protected area in a move that raised fears that outsiders would manage the process (Bryant 2000; Lawrence 2002).

The PAFID thus had to battle on several fronts. Employees used various tactics to press the government, while ever attentive to the possibility of using a good name to advantage. There was some success in the case of illegal logging. While community organizer Ruel Belen documented Tagbanua complaints to pursue the matter through regular channels, senior staff and board members used personal contacts to lobby DENR managers. A key objective was to ensure that any investigation was removed from a local government seen as biased against the Tagbanua. This objective was achieved in 1996 when DENR Secretary Victor Ramos announced that senior management would investigate the matter in a move that signaled top-level opposition to continued local logging (Austria 1997). This victory was due in no small measure to the PAFID's "high credibility" with senior DENR staff such as Delfin Ganapin and Tony La Viña, who were keen to use the pressure of a respected NGO to force change (Ganapin 1997; La Viña 1997).

The battle against tourism was more difficult. Here, the PAFID was unable to use moral capital linked to the DENR as leverage since tourism was the preserve of various agencies *and* enjoyed top-level support. Belen (1996) wrote to Tourism Secretary Mina Gabor, President Fidel Ramos, and others in July 1996 to forward a TFCI petition asking that Coron Island be exempted from "poorly planned tourism development that remains insensitive to their culture." This effort did not work. Secretary Gabor (1996) replied that due to its "unique natural attractions" including "the only well-preserved limestone forest in the country," the department was "seriously studying how tourism should be developed in this area" (see also Honasan 1996; Tan 1996).

The conservation scheme was the most difficult case of all. One tactic was to participate in the community consultation of the National Integrated Protected Areas Program (NIPAP) held at Coron Island in November 1996. The PAFID adopted a dual strategy when that consultation simply sharpened the divide between the Tagbanua and NIPAP. Thus De Vera and colleagues pressed counterparts at the DENR and NIPAP to reexamine how the initiative addressed community concerns. They emphasized that local opposition reflected a feared "loss of control" that might endanger Tagbanua livelihoods. This pitch was only partly successful. The NIPAP did seek to allay fears when it pressed the DENR to increase Tagbanua representation on the proposed local PAMB, yet the NIPAP professed surprise at "local antagonism" that was put down to "misinformation" derived from the actions of the PAFID (NIPAP 1997: 4; Calanog 1997; De Castro 1997; Lawrence 2002). As a result, NIPAP Philippine Co-Director Lope Calanog (1997) looked for ways to "try and win them to its side" through promises of local employment (also De Castro 1997).

The PAFID meanwhile redoubled efforts to win a CADC prior to the area's confirmation as a protected area. Staff pointed out that a CADC would enable the TFCI to promote local conservation. This was because it would delegate power to those "best placed" to do the work—the islanders themselves. NGO staff also noted the strong legal case for a CADC. Here, De Vera relied on board member and legal activist Donna Gasgonia (1997), who, in a paper prepared in 1997, argued that the claim was guaranteed under the 1987 constitution. This paper went to DENR Undersecretary La Viña, who confirmed this view in early 1998 (PAFID 1998).

Still, there was an ambiguous outcome in the end. Patient lobbying by the PAFID paid off when the Tagbanua were awarded a Certificate of Ancestral Lands and Waters in July 1998 (Belen 1998). The size of the territory covered (22,284 hectares) represented a triumph for the TFCI and the PAFID. Yet opponents appealed to the president to reduce this area, suggesting that the battle was not over. Matters were complicated in 1999 when NIPAP's Calanog signed a memorandum with a barangay captain on behalf of a majority of Tagbanua residents affirming protected area status (Verian 1998; NIPAP 1999). Coron Island was thus the object of rival designations, setting the scene for tough negotiations over management (Lawrence 2002).

What are we to make of the political influence enjoyed by the PAFID in this case? Advances were certainly made—notably with regard to illegal logging and the award of the Certificate of Ancestral Land and Waters. There were nonetheless setbacks too. The NIPAP scheme went ahead slightly modified but with basic parameters largely unchanged. Tourism plans for the area were not stopped. Top-level political support and rival political interests can thus limit what even a well-respected NGO can achieve in specific policy battles.

At a broader level, this example suggests that there is a pressing need for NGOs to work together and with other allies to reform state structures resistant to change if moral capital accumulated as a result of routine interaction is to result in substantial "pay-offs" through systematic policy influence (Alegre 1996; Sidel 1999). The campaign to "green" the state according to commitments made at the 1992 Earth Summit is a case in point. True, the rhetoric was enough to make even the most jaundiced NGO veteran wince. President Ramos thus pronounced the Philippine state a "pioneer" when it created the Philippine Council for Sustainable Development (PCSD) in September 1992. The PCSD had thereafter become "a truly participatory decision-making process," and its main output, *Philippine Agenda 21* (PA21), was presented in "a spirit of unity, solidarity and convergence" (PCSD 1997: iii).

Some activists sought to use official rhetoric as a lever with which to advance reform through the PCSD. The Haribon Foundation was one such NGO. As leader Cristi Nozawa (1996) noted, they felt that PA21 indicators of social and environmental "progress" could help bolster the position of reformers in the DENR even as membership in the PCSD was a way of "getting ammunition" to strengthen criticism of official practice. Other NGOs expressed skepticism. As PRRM leader Horacio Morales remarked, the absence of "time-bound plans" meant that it was exceedingly difficult if not impossible to translate indicators into specific deliverables. The fear was that this process was simply "cosmetic environmentalism" (Morales 1997).

Yet routine interaction is about more than simply acquiring moral capital with state partners in order to gain policy influence. It is also a potentially efficacious method of boosting a reputation with other actors. Local communities, for instance, can be keen to work with NGOs that "get things done" on their behalf. For the PAFID, deliverables revolve around community organizing and official land entitlement. For Haribon, expectations center on community organizing, livelihood development, and paralegal training. Effectiveness in actually delivering the goods bears some relation to how NGOs are seen locally. As discussed below, the fate of constructive engagement is usually predicated on the assumption that NGOs work with official counterparts. Thus routine interaction with the post-Marcos state is often a precondition for an NGO to accumulate moral capital with community partners.

The same can be said about NGO relations with donors. Most donors have been keen for NGOs to work with Aquino and her successors to strengthen democratization in the Philippines. In the case of USAID, for example, NGOs were encouraged to "build cooperative arrangements" with government after 1986 (Magno 1997). As part of the U.S. government, this agency has long played a role in Philippine life. With the fall of Marcos and rising anti-U.S. sentiment, USAID was anxious to safeguard America's regional role by helping to consolidate formal democratization even as this process would subvert efforts for more radical political change (Putzel 1992; Clarke 1998). Support for NGO–People's Organizations (PO) partnerships designed "to build a constituency" for policy reform in the 1990s continued this process (USAID 1996: 2; Magno 1997). Here too then, routine interaction between NGOs and state agencies was important to accumulating moral capital with donors (a similar story in the case of the Netherlands is related by Teunissen 1997).

Working with the post Marcos state has therefore provided an opportunity to accumulate moral capital with those appreciative of NGO "pragmatism" and "responsibility." When NGOs strive "to put their best foot forward" through carefully packaged annual reports, funding applications, or position papers, they highlight "good practices" through routine interaction. Clear if modest benefits flow from this approach. However, it was also highlighted that there are limits here that implicate the quest for moral capital itself. Hence many NGOs often lace routine interaction with criticism of state partners as part of their critical engagement strategy.

Maintaining a Critical Distance

A capacity for autonomous action is usually seen to be synonymous with an ability to criticize "deficient" policy. Thus NGOs are important because they are "public watchdogs" who by "being vigilant keep government on its toes" (Vitug 1997). In this view, the NGO that is so enmeshed in routine interaction that it does not fulfill this function is a lesser organization. Criticism of state practices is seemingly part and parcel of what it means to be an NGO with moral capital. Indeed, criticism facilitates the accumulation process. That NGOs are *expected* to criticize state partners periodically reflects especially the view of them as moral agents discussed earlier. Much moral concern with which NGOs are linked relates to perceived deficiencies in official practice. Thus they are expected to criticize because of a gap between what state agencies do and what NGOs (and others) wish them to do.

While NGOs "make trouble," they must not become mere "troublemakers" if they wish to maintain a multifaceted accumulation strategy. They must pay close attention to *which* state practices they criticize as well as *how* they criticize. There is a need here, then, to avoid being seen as "making a living" out of state condemnation through an "advocacy as antagonism approach" (Austria 1997). As such, critical interventions require careful planning. How will criticism play with donors, the media, a middle-class public, or community groups? Will a reputation be enhanced and, if so, in what ways? How will the target of criticism react?

Much strategizing in the decision to criticize state partners reflects a *rule of specificity* that suggests that NGOs will avoid blanket condemnation of them. The *degree* of specificity can range from an entire policy to specific implementation issues even as resort to this rule is generally indispensable to critical engagement. This is because it is a means to reduce the chance that criticism will endanger relations with state partners—perhaps even leading to an NGO's being labeled as "communist" (Clarke 1998). The rule of specificity makes sense when it is recognized that the need to oppose is matched by the desire to cooperate. It is a rough and ready "cost-benefit" mechanism whereby NGOs may seek to maximize accumulation through criticism without jeopardizing the ability to accumulate through routine interaction.

How does it work in practice? Sometimes, there is compunction to oppose an entire policy because it is seen to lead to social or environmental degradation. Here, the rule is stretched to the limit and containment of conflict is difficult. Still, there is an effort to stay focused on the policy by asserting what is wrong with it as well as why incremental change is not feasible. The challenges here are illustrated in the anti-logging campaign that dominated environmental politics in the late 1980s and early 1990s. This campaign was coordinated by an NGO coalition, the Task Force Total Commercial Log Ban (TFTCLB).

The Haribon Foundation played the lead role in the TFTCLB as the coalition campaigned for a total ban on commercial logging. Much of their success was due to a strategy of focusing public attention on deforestation as an example of policy failure. The NGOs first needed to "frame" deforestation as a national problem. Haribon had acquired experience here in running the campaign to "save" Palawan's forests. That regional campaign generated media attention that had indirectly highlighted the general plight of the national forests. Haribon and the TFTCLB thereafter amplified the message by endless repetition of statistics. On the eve of Senate deliberations in July 1990, for example, the coalition noted that "in just 40 years" forests were reduced from "75% of our total land area" to 20 percent in 1990—even though the country needed "54% forest cover" to protect "the integrity of the soil, to prevent its erosion, lowland flooding, river siltation and corals suffocation" (Haribon Foundation 1990c: 4; see also Repetto and Gillis 1988). Cover was needed to "insure water supply" and to ensure that habitat was safeguarded for "rare tropical flora and fauna" (Haribon Foundation 1990c: 4).

Deforestation was linked to policy failure by successive governments. Decades of unfettered logging reflected a system that enriched license holders at the expense of the nation. As Kalaw (1990: 8) noted, there was "an income estimated at US$42 billion" that accrued to "480 logging concessionaires" even as it "contributed directly to the poverty of 18 million people living in the Philippine uplands." What is more, concessionaires failed "to re-forest the areas which they have logged, destroying, in the process, the fragile tropical ecosystems" (Kalaw 1990: 8). While concessionaires were greedy, the problem was "essentially an institutional one, having to do with rules of access and control" whereby "big and influential concerns" triumph over "small-time operators or community interests" even as concessionaires "make a killing" in rents at the expense of the public exchequer (Kalaw 1997: 19; see also Kummer 1992; Broad with Cavanagh 1993; Vitug 1993; Ross 2001).

Once policy failure and deforestation were thus connected, debate shifted to what *new* policies were needed. The Haribon-led campaign fought the position supported by industry, pro-logging congressional leaders like Senator Heherson Alvares, and DENR Secretary Factoran that called for a selective log ban. Criticism reiterated the dire condition of the nation's forests and the concessionaire role in overlogging. The selective log ban would be like "a balding man who is trying to regenerate his hair but is pulling off those which has [*sic*] been left" (Haribon Foundation 1990c: 4).

The coalition mostly stuck to the policy in question throughout the debate. Still, the manner in which Kalaw publicized policy "failure" in national and international media angered DENR officials. The nadir came in autumn 1990 when he was associated with a call for the DENR secretary to resign. Relations with the DENR had been good until then; he had even received a DENR award for "espousing the cause of social equity in resource use and the significance of the role of non-government organizations and rural communities in sustainable development" (Haribon Foundation 1990b). Yet, as differences grew between Kalaw and both Secretary Factoran and Undersecretary Roque, conflict spilled over into other areas (Factoran 1997). As we will see, such antagonism was behind Haribon's departure from the debt-for-nature swap in which it and the DENR were involved. Haribon's experience with the anti-logging campaign, then, highlights possible limits to the rule of specificity as a means to pigeonholing conflict between an NGO and a state partner.

More often than not, NGO criticism focuses on policy implementation. This is hardly surprising, since it is only as policies are implemented locally that communities experience the costs and benefits of state action. Costs have been substantial, as community rights and supporting ecosystems have been sacrificed to aid development. This illustrates the flawed nature of the policy process, since "environmental policy and legislation in the Philippines are probably comparable to the best in the world," though "all these policies, laws and programs . . . have failed to adequately protect our environment in the past two decades" (Serrano 1994: 4–5). Further, NGOs devote much energy to work in communities and are thus well placed to offer criticism from this vantage. This aspect of NGO life has ramifications for the quest for moral capital. Many expect criticism of state policies to emanate from a solid grounding in community realities. Even state partners may be receptive to criticism when it is seen to reflect such expertise. The rule of specificity here resonates therefore with a perception of the sort of criticism that is "valid" for state actors.

The Bolinao conflict that erupted in the 1990s over plans to build a cement plant at the expense of local coral reefs is an example of criticism targeting implementation. This case also enables us to consider how criticism is communicated. There is evidence here, too, of criticism voiced in a moral idiom. Finally, this conflict shows that focused criticism can facilitate moral capital accumulation.

Haribon became involved at Bolinao due to its expertise on coastal resource management (Lee 2004). Funded by the Food and Agriculture Organization and the International Development Research Center, it joined scientists from the University of the Philippines in Manila in a program of research, community organizing, and livelihood development. Haribon's main brief was community organizing, and four employees established themselves locally to this end in 1993. Much work needed to be done. The reefs were deemed among the best nationally, and this biodiversity "hotspot" was a vital resource to municipalities around the Lingayen Gulf. Yet poor fishing practices were destroying them. Haribon staff confronted this situation by helping to found people's organizations in the area through community meetings and workshops. These organizations were trained to promote sustainable fishing techniques and other environment-friendly activities such as handicrafts production. Haribon's legal arm, the Tanggol Kalikasan, provided paralegal training to community leaders so that they could stop illegal fishers from entering the area.

These activities alienated barangay captains involved in the fishing industry. Not surprisingly, a petition against this industry by one people's organization fell on deaf ears in local government. Tension increased when plans for a multimillion-dollar cement plant were announced. This plant was to be built by the Taiwanese-funded Philippine Cement Corporation and would include twenty-two quarries, a sixty-megawatt coal-fired power plant, a ten-kilometer conveyor belt, and a 550-meter wharf. While the community would gain jobs overall, and the government would receive six to eight million dollars (albeit after a five-year tax holiday), activists pointed out the costs. There would be "a devastated landscape" and "the obliteration of marine habitats," leading to long-term economic collapse and "poor health and misery" for residents (Doyo 1997: 9). As the controversy heated up, local leaders attacked Haribon. Staff recalled how their work was slated as "anti-development" even as they were branded as "communists" in the local media (Burillo 1996; Mesa 1996; Turion 1996).

These attacks strengthened the bond between Haribon employees and the communities in which they worked. By staying put, staff earned the respect of residents hitherto skeptical of NGO motives based on prior experience. The attacks also prompted Manila to make this an issue for national action by the organization. Haribon was not alone in this struggle. The Movement of Bolinao Concerned Citizens was at the center of the protests, with technical and logistical input provided by Haribon and University of the Philippines staff. Yet Haribon's local track record as well as its national and international reputation for community work and ecological research meant that it was ideally placed to contribute to this issue (Supetran 1997).

Haribon criticism accounted for tough political conditions surrounding the proposal. In addition to local government support, political leaders including President Ramos backed the proposal. Yet Ramos also wished to be seen as an environmental president. Since the proposal required an environmental compliance certificate from the Environmental Management Bureau of the DENR, here was a chance to test that "commitment." Haribon efforts to pressure the DENR and the president followed two paths. Staff joined local and national protests designed to ensure that the case received media coverage. The conflict was featured in the *Haribon Quarterly* distributed to members (Haribon Foundation 1995; Albano-Vitug 1996). In this way staff enhanced the turnout at events such as the 1995 protests at the DENR headquarters. Staff also supplied information to journalists at the *Philippine Daily Inquirer*—a center of media opposition to the plant (de Quiros 1996a, 1996b; Doyo 1997; *Philippine Daily Inquirer* editorial 1996).

Meanwhile, senior staff pursued quiet diplomacy with top politicians and bureaucrats. A key intervention was a confidential letter from Kalaw to President Ramos on January 23, 1996. In it, Kalaw summarized opposition to the proposal, referred to Ramos's environmental commitments, and suggested that the plant be moved elsewhere. The tone of the letter was "reasonable" and reiterated the ethical importance of conservation. It was also constructive, in that it proposed a solution that would enable the government to protect the reefs *and* to build the plant. The response in a letter from DENR Undersecretary La Viña was conciliatory, highlighting the "positive" suggestion about an alternative site contained in Kalaw's letter. In the end, the government withheld the environmental compliance certificate, thereby handing a victory to the campaign in August 1996. This case illustrates how an NGO uses various means to criticize policy implementation. The selection and timing of those interventions are not haphazard. Thus there was much discussion about Bolinao between the board and senior staff as well as between the NGO and other organizations (Haribon board minutes 1994–96).

For a reform-minded NGO like Haribon, getting it right here was important because more was at stake than simply a decision over a cement plant. It was also about how partners would view criticism of the state in this situation (including whether such criticism was sufficient). Getting it right was thus about making a stand without alienating state and donor partners. Adopting a "responsible" tone was critical, even as the suggestion of an alternative site for the plant was particularly inspired. The latter signaled to the government that Haribon appreciated official constraints and that it wanted a solution acceptable to all. It enabled the government to save face. Skilled work in public, meanwhile, provided the political pressure needed to convince Ramos to take the controversy seriously. Haribon's good name in the reform-minded media was an asset, as the NGO could put the case against the plant to a wide audience (Barrera 1990; Nuyda 1997).

Along with the fate of Bolinao's reefs, then, Haribon's reputation was on the line in this campaign. That the NGO got it just about right is to be seen not only in a successful campaign but also in acknowledgment by government counterparts. As Environmental Management Bureau chief Amelia Supetran (1997) noted, Haribon was "one of the more serious NGOs" involved and played a "key role." The NGO was, added DENR Undersecretary and former NGO leader La Viña (1997), instrumental in contributing to the "political compromise" that ensured that, against the odds, the plant was not built. Here, then, is a case in which an NGO interweaves focused criticism and routine interaction. Indeed, while the act of criticism is presented in a nonthreatening way, it nonetheless remains faithful to an organizational mission supportive of community-based sustainable resource management.

The critical engagement analyzed above seems to address many of the needs of reform-minded NGOs in the post-Marcos era. It does not eliminate the ambiguity at the heart of state-NGO relations but rather seeks to turn that ambiguity to advantage through a multifaceted political strategy. By switching between cooperation and criticism NGOs can maximize the chance that moral capital will be accumulated even as policy change is encouraged. If much attention has been given thus far to NGO strategies vis-à-vis state agencies, this is because of the importance of the Philippine state to what many NGOs do. And yet this relationship must not obscure the political strategies that are developed in relation to others.

Cultivating the Grassroots

Community support for an NGO can never be taken for granted. Political, economic, and cultural divisions often render pan-community endeavors problematic. Other obstacles include local distrust, skepticism, and hostility toward outsiders. NGO practices designed to overcome these hurdles and facilitate work in communities are here labeled constructive engagement.

NGOs aim to build relationships through activities ranging from the creation of people's organizations to the promotion of alternative livelihoods. Two things happen here at the same time. There is, first, an effort by NGOs to promote practices consonant with their visions and missions. These practices are often in conflict, though, with other practices linked to personal survival or political intimidation. Second, there is a campaign by NGOs to convince potential partners of their local utility. Why should residents work with them? Can NGOs keep their promises? Will there be long-term support to the community? These questions indicate that community support must be earned even as they alert us to the actions NGOs need to take in order to build moral capital locally.

Promoting constructive engagement with communities can be difficult, given diverse community settings. There are also difficulties in gauging how community members react to NGOs. Yet my goal here is not to specify what residents think about an NGO but to consider possible patterns in what NGO staff say and do locally. It is strategic thinking and behavior by NGOs—and its link to moral capital—that is at issue here. The PAFID's work at Coron Island illustrates how connections that have developed over time can lead to a durable relationship. The nature of the first contact was critical. In 1985, Coron Island barangay councilor Rodolfo "Codol" Aguilar approached the PAFID after learning that it had helped another local community to obtain a DENR Community Forest Stewardship Agreement (CFSA). Though not perfect, this precursor to the CADC seemed one means for islanders to repel unwanted development. The NGO was initially invited to Coron to explain the scheme, and after much debate, a majority of residents supported the CFSA. With a letter of invitation from the Tagbanua, PAFID staff thereafter helped create the Tagbanua Foundation of Coron Island (TFCI) and prepare the CFSA application. The combination of the award of the CFSA in 1990 and the TFCI's maturation under Aguilar meant that residents were able to reduce local development intrusions (Jaravelo and Tolentino 1989; PAFID 1998).

Yet TFCI authority required constant assertion. The PAFID helped develop a management plan and taught residents how to cultivate cashews for the market. Outsiders, meanwhile, once more threatened Tagbanua control. Tourist operators took tourists to the lakes, beaches, and caves that were, in the first case, a source of income (edible birds nests), in the second, water, and, in the third, burial sites. Individuals with government connections felled trees in violation of the CFSA. Coron Island was proposed as a site for the European Union's NIPAP scheme, with backing from the DENR and NGOs such as Conservation International. Migrant fishers and even some community members used dynamite or sodium cyanide to catch fish, thereby destroying the coral reef.

Residents turned again to the PAFID for help. The NGO had won kudos for its help in creating the TFCI and winning the CFSA. Subsequent ties were further proof of local commitment. This record was important in the organization's next phase of intensive local involvement in the mid-1990s. The PAFID agreed to help out once a letter of invitation from the residents was received. The importance of such a letter to the NGO's strategizing is not to be gainsaid, by the way. It provides staff with an effective response to charges of meddling, and it insulates the NGO to some extent against intracommunity divisions. It also strengthens the organization in the eyes of donors keen to see community links. Finally, the letter makes the symbolic but important point that the community is formally in charge of joint projects, thereby according that community respect. Politically astute, the letter of invitation is also a testament to the reputation of an NGO such as the PAFID. It may even illustrate how moral capital accumulated in one community may lead to its acquisition in other communities.

If constructive engagement involves the generation of measurable outcomes, it can also be about capacity building and knowledge dissemination. These processes may defy measurement but are nonetheless important. Indeed, they can prompt changes in community perspectives with implications for resistance and policy implementation. As such, it is important to consider this aspect of NGO work. PAFID involvement at Coron in the mid-1990s is illustrative here. In particular, the confrontation over the NIPAP emphasized the multifaceted behind-the-scenes influence of the PAFID vis-à-vis the Tagbanua. It also becomes clear how the generation of moral capital can gain momentum through ongoing cooperation.

Residents were uncertain about the NIPAP. True, the scheme was about something dear to them—the protection of Coron Island. Yet suspicion of official initiatives left many uneasy. Could proponents be trusted to account for Tagbanua concerns? How would NIPAP management affect existing practices? How might acceptance of the NIPAP affect the quest for ancestral domain? The PAFID played a key role in assisting them on these questions. It did so notably by helping the TFCI to prepare a two-day NIPAP consultation held on Coron Island in October 1996. The meeting was designed to explain the NIPAP to residents. For NIPAP Director Lope Calanog, the point was "to pave an entrance" to this hot spot by building local rapport, primarily by explaining local "benefits" (Calanog 1997; see also Lawrence 2002). That the consultation turned out quite differently was due notably to PAFID influence.

The NGO's employees were already skeptical about the NIPAP. They were familiar with it through work in Mindoro, where indigenous people rejected the NIPAP for fear that designation would undermine ancestral claims (NIPAP 1996; Rood 1998). Hence Dave De Vera advised the TFCI to compile a list of local practices for a Tagbanua ancestral law. He suggested this step would bolster community pride, since the law profiled Tagbanua skills even as it would be a basis for a CADC. The law also provided the TFCI with its own proposal to present at the NIPAP consultation (TFCI 1996; PAFID 1998).

The law was prepared by the TFCI with PAFID assistance. The process involved distilling oral traditions into a written account of locally "sustainable" land and water management. Logistical and intellectual support from the PAFID was vital. Community organizer Ruel Belen spent much of October on this. Not only did he transcribe the oral account into a manuscript; he also translated what was agreed upon into legal idiom appropriate for official consideration. Much depended therefore on this employee's precision. Yet he was highly respected and trusted by a Tagbanua leadership accustomed to working with PAFID employees who "delivered the goods." As a result, preparations went ahead smoothly and quickly.

Other staff came to help as the date of the meeting neared. That two experienced policy analysts as well as leader De Vera turned up underscored the PAFID commitment. They assisted Belen and the TFCI with community validation of the law and offered advice on strengthening it. They also made crucial interventions during the consultation itself. Thus, while the TFCI chaired the consultation, PAFID employees provided support from the sidelines. Prior to the start of the meeting, for example, De Vera met with Calanog to discuss Tagbanua problems with the NIPAP—especially the lack of Tagbanua control of the proposed PAMB—to ascertain where NIPAP stood on this matter. Staff helped Aquilar thereafter to ensure that audience participation reflected concerns about the protected area. Above all, organizing work paid off when the NIPAP team failed to intimidate the Tagbanua. Local self-confidence was especially demonstrated on the second day when the TFCI completed their critique of the NIPAP proposal and presented their own ancestral law without hitch and without PAFID staff in attendance.

What to make of PAFID's role here? The NGO was central to the process of articulating a Tagbanua perspective. While the ancestral law reflected the views of the Tagabanua, the PAFID helped them to see how it could be used strategically to promote ancestral domain over the NIPAP scheme. Yet the stakes were high: one visitor even suggested that Tagbanua would "no longer be poor" if they accepted the NIPAP and "this luck might never come their way again."[1] However, because they had faith in the PAFID—and believed it to be acting in the Tagbanua's best interest—they rejected such blandishments in favor of action suggested by the PAFID.

Such trust reflected bonds of friendship developed with PAFID staff over the years. In the case of Belen, local acceptance was solidified through residency on the Island. Yet local acceptance was due to more than the activities of this one employee. Since various staff had visited Coron since 1985, trust and friendship thus extended to others as well. Further, this process did not happen automatically but reflected conscious effort. A staff member summarized one visit: "Our first activity was to establish immediate rapport with the community" (Jaravelo and Tolentino 1989: 1). This effort took various forms. Thus rapport was established when PAFID employees were able "to mobilize at least three boatloads of people to join us when we went for a bath at a hot spring beside the seashore" (Jaravelo and Tolentino 1989: 1). To nurture connections in this way was vital to constructive engagement between the PAFID and the Tagbanua. In fact *how* staff conducted themselves with community members was often as important as *what* project goals they achieved. This is not surprising, since communities like the Tagbanua are beset by elites disdainful of local people (Rood 1998; Bryant 2002a; Lawrence 2002) and they often demand respectful conduct from NGOs.

The experience of the Haribon Foundation at Mount Isarog provides further evidence to this effect. It also alerts us to the manner in which constraints on an NGO can lead to the devaluation of its local reputation. As noted, Haribon became involved here in 1989 when the student chapter did a park survey. Systematic work began when funding was obtained thereafter. Five barangay were selected based on criteria linked to human settlement in the park, extent of environmental destruction, road accessibility, provision of government services, and the peace and order situation (the NPA were locally active). Employees assigned to each barangay were responsible for community organizing, alternative livelihoods, and research. Haribon entered these districts without a letter of invitation. Employees instead held "house-to-house conferences" at which they introduced themselves and explained what biodiversity conservation was and why it was needed. It was also when they set out the community-based management approach that they wished to see introduced. This method of introduction was labor intensive but vital if the NGO was to make inroads into these communities.

Local circumstances were not propitious since many communities were dependent on illegal logging. As the leader of the Cawaynan people's organization recalled, when the NGO entered the barangay people conducted "illegal logging, *kaingin* [shifting cultivation], and other forest extraction activities as the major source of their livelihoods" (Vale 1996). Members of the local government were involved in illegal logging and bitterly opposed Haribon's arrival. While the NGO was able to organize in Cawaynan, the leaders of that people's organization were thereafter locked "in conflict with the local government" (Vale 1996). In Panicuason, Haribon fared even worse as implacable local opposition torpedoed a people's organization. Endemic poverty was also a factor. Many people were dependent on landlords who were indifferent to the Haribon message. In Lugsad, 90 percent of the 160 families were tenants, growing crops such as sugarcane and also reliant on forest extraction. There was little room for maneuver here. Appeals to them as to the merits of conservation—the protection of biological diversity and a move away from forest extraction—thus fell on deaf ears at first.

Progress was nonetheless made. People's organizations were created with each organization comprising between twenty and thirty members. These organizations bore the stamp of the NGO, inasmuch as they featured identical committees on livelihood, membership, finance, and paralegal concerns. The day-to-day activities of the committees reflected the input of some residents as the latter sought to advance their interests through the new institutions. Here too, Haribon was vital. Its staff provided leadership training and accounting skills to enable the organizations to function. They offered paralegal training to facilitate the fight against loggers even as employees taught ecology and agroforestry to promote alternative livelihoods.

Haribon earned local respect and support through these activities. The manner in which it entered the community was important since house-to-house conferences helped employees learn about local concerns in a respectful manner. The alternative livelihood scheme in particular signaled that the NGO understood the need to combine talk of conservation with practical steps to promote local livelihoods. The focus was on agroforestry, with each people's organization given funds to develop seed nurseries and demonstration farms. The nurseries would increase the supply of woody and fruit tree species so that people involved in logging could safely abandon that practice. The demonstration farms would do likewise but would also illustrate sustainable farming methods designed to appeal to tenants and landlords.

The scheme had a negligible impact on livelihoods. A typhoon ripped through the area destroying seedlings and devastating crops. The modest nature of the scheme itself meant that it could only have had a minor impact. On the one hand, there was "easy money" from logging. One resident observed, "If the project can guarantee livelihood [in contrast to logging] I support it, but I doubt it" (Riipinen 1995: 55). On the other hand, there was dependency on landlords not in the project. The head of the Lugsad people's organization had to clear forest owned by his landlord for a sugarcane plantation but could not challenge him for fear of dismissal (Perez 1996).

Haribon still won local support. In Lugsad, residents appreciated the provision of education on sustainable farming and agroforestry. One participant explained, "I have learned much on new planting techniques and forest conservation through this project" (cited in Riipinen 1995: 54). The leader of the Cawaynan people's organization reflected a general view in observing that residents "now understood the value of protecting their forest" against clear felling it for immediate gain thanks to the hard work of Haribon staff (Vale 1996). On the one side, there was a fuller recognition that valuable biodiversity was being lost. On the other, there was the prospect of new economic activities predicated on *in situ* conservation. Recognition of the virtues of the Haribon project was neatly summarized by one villager: "This does not help us to change everything but it helps us to get some more income to the organization, and through this we can get some benefits to us, as well as to the environment" (cited in Riipinen 1995: 53–54).

Haribon used various means to communicate its message. Day-to-day project activities were the main method, along with house-to-house conferences. Novel tactics were also used. For example, the NGO distributed forest conservation comics that it wrote and translated into the local dialect. These activities appear modest when compared with the larger realities of poverty locally, yet staff dedication clearly inspired some to take up the conservation struggle. This process facilitated the accumulation of moral capital in the communities. In Cawaynan, it was noted that the Haribon message was gaining ground even as illegal logging declined (Vale 1996; Riipinen 1995). Indeed, the experience of this people's organization as a "protector of the environment" was becoming known externally (Vale 1996). Thus it was profiled in the local newspaper as well as in Haribon's own national newsletter (Navarro 1995).

Haribon earned kudos by joint action against illegal loggers—a dangerous practice that risked violence. Yet such action was an "integral part of Haribon's work" (Luna 1996). For example, staff helped the Lugsad organization erect a human barricade across a road used by loggers, thus stopping a shipment, and thereafter logging was greatly reduced in the community. Success here also gave an enormous boost to the people's organization since it demonstrated their ability to stop loggers. Finally, it was a boon to Haribon's reputation as it demonstrated the value of its work and a willingness to show solidarity with local partners even under duress. These interventions did not win the battle against illegal logging at Mount Isarog, though their cumulative effect did contribute to its decline. Reporters noted this impact in assessing Haribon's role in the struggle. Thus Doris Gaskell Nuyda (1997), writing in the *Philippine Daily Inquirer,* described the decline in logging as a "success story" in the national environmental battle in which local groups "contained" the threat with firm support from Haribon.

Yet Mount Isarog was no clear-cut success for Haribon. Funding was a headache. The NGO suffered a severe budget cut locally in 1994 that led to the loss of eight of ten project employees. Project success was jeopardized since, as project officer Noel Resurrecion (1996) observed, they "pulled out too soon" from the communities. While external forces beyond Haribon's control were to blame here, the pullout nonetheless generated tension between staff and the local organizations. One leader (anonymously) explained that he and his colleagues experienced difficulty "ever since the Haribon Foundation has left them on their own." Indeed, he added, it had "left them too early" as they were not ready "to take organizational responsibilities all by themselves." There was resentment too that only six months' notice was given and no extension was granted despite local appeals. Moreover, there were local knock-on effects. The pullout triggered a decline in local interest in the people's organization, in that residents had often looked to Haribon for guidance. True, the impact varied between barangay. In Lugsad, the people's organization flourished, since a micro-finance project was initiated with Haribon assistance instead. Also, this group enjoyed strong leadership from residents who doubled as district officials. That said, even here, one leader confided there were "organizational problems" following the pullout.

Haribon staff worked hard to contain the fallout as manifested in declining local moral capital. They knew that their name was sullied even though they were not to blame for the budget cuts. However, NGOs often face local frustration at such times due to an expectation that staff can fix problems (Alegre 1996; Fowler 1997). Haribon thus pursued several avenues to repair the damage. They persuaded the four people's organizations to form an umbrella group, the Anduyog Federation, to fill the gap left by Haribon even as staff helped it to win funding. A micro-lending project was followed by a capacity-building grant from the Foundation for the Philippine Environment related to community organizing, seed propagation, lobbying, and fund-raising. These actions softened the blow of the pullout. They also underscored Haribon's ongoing commitment to local partners even in the face of budget cuts (Resurrecion 1996). Representatives of the local groups, meanwhile, were "trying to understand the situation" and make do under the circumstances (anonymous interview).

Constructive engagement is not an easy process. Community support must be earned through interaction in which NGOs demonstrate competency and solidarity. This is not straightforward even under ideal conditions. Support often wanes when NGOs do not keep promises. The accumulation of moral capital here, then, requires skill as links are cultivated and trust is established. A multifaceted quest for moral capital leads NGOs into differentiated political strategies as they engage with other actors such as state agencies and local communities. Complex political "artistry" is called for as organizations negotiate relationships in promoting visions and missions. This process is partly a matter of reputation. To accumulate moral capital is inescapably a matter of being attentive to shifting political opportunities. At the heart of this process, though, is sensitivity to what is "expected" of an NGO—without which an organization is liable to suffer disappointment in pursuing social change.

CHAPTER 5

Financing Prophets

Funding is widely seen as the Achilles heel of NGOs inasmuch as the need for cash can lead them into dubious tradeoffs at odds with their mission. They sometimes become, in the title of one work, "too close for comfort" to donors and to government (Hulme and Edwards 1997a). There is a paradox here. For many, one of the strengths of NGOs is their not-for-profit status because it appears to reinforce perceived moral and altruistic qualities (Korten 1990; Sogge 1996a; Slim 1997). Increasingly, though, this status is seen as a source of weakness. Being a not-for-profit operation seemingly condemns NGOs to perpetual economic insecurity. By eschewing profit, they render the task of financing their chosen role difficult. It dooms them to a life of continual fund-raising (Bennett and Gibbs 1996; Weisbrod 1998; Fowler 2000). Shifting economic conditions appear on balance to add to this picture of misery. True, in the Philippines the surge of funding in the late 1980s and early 1990s was a boon for many NGOs, but by the mid-1990s donors began to switch funding from a "prospering" Philippines to supporting "needier" countries such as Vietnam (Gray 1999). The point here is that the funding "roller coaster" is fickle, rendering NGO financing unstable.

Clearly, NGOs need sophisticated strategies if they are to acquire the money needed to sustain themselves. Yet these strategies are not decided in a vacuum. How do organizations solicit funds? From whom do they accept money? What are the conditions under which financial assistance is accepted or sometimes rejected? These questions raise issues about identity. They also point to dilemmas and opportunities that NGOs confront through integrating financial imperatives with concerns and perceptions about morality and altruism. The search for funding is thus associated with the quest for moral capital. How NGOs accumulate financial resources may tell us much about how they go about building reputation (and vice versa).

Money Matters

That money plays a pivotal role in the operation of nonprofits is a recurring theme in the literature. The general message is that money matters render nonprofit actions ambiguous, as core visions and missions may be compromised (Smillie 1995; Hulme and Edwards 1997a; Anheier and Salamon 1998). As Weisbrod (1998: 12) observes, the central dilemma facing nonprofits is "how to balance pursuit of their social missions with financial constraints when additional resources may be available from sources that might distort mission."

For some, the problem is donor dependency. Work by resource mobilization theorists, for instance, highlights the impact of the dependence of social movement organizations on members of an elite for funds through philanthropic foundations, churches, and government agencies. McCarthy and Zald (1987: 385) remark that such dependency has the effect of "directing organized dissent into legitimate channels," thereby "diffusing the radical possibilities of dissent in general" (cf. Mercer 2002). Nonprofits may even prioritize donor utility over that of those they wish to help. This fear is echoed in suggestions that NGOs are "too close for comfort" to donors (Hulme and Edwards 1997a).

There is also the view that sees nonprofits as niche players in the market. Here, the study of these organizations mainly by economists leads to a functional view. One writer opines that the sector develops due to "constraints on governmental provision of collective goods" and thus supplements such provision "helping to meet the remaining, undersatisfied consumer demand" (Weisbrod 1977: 15). Meyer (1995: 1279) argues that NGOs reflect "entrepreneurial activity" and are "producers of international public goods," such as biodiversity inventories in Costa Rica and debt-for-nature swaps in Ecuador. True, being a niche player does not necessarily lead a nonprofit to diverge from its mission (which, in any event, can shift over time for other reasons). And yet, the close links that tend to develop with elites in business or government as part of this agenda does, she concedes, "alter the incentives that nonprofit NGOs face [and] the allocation of benefits of such NGO activity cannot be expected to be either optimal or egalitarian" (Meyer 1995: 1285).

Finally, authors highlight how the need to raise funds can lead nonprofits into becoming more like the for-profit sector. Commercial activities are seen to be the way to boost income in lieu of increased donations. The search for salable goods and services has involved nonprofits in thrift shops, real estate, greeting cards, and consulting. As Weisbrod (1998: 9) observes, nonprofits thus pursue revenue "in the same ways that private firms do," and by "becoming more like them may undermine the fundamental justification for their own special social and economic role." Such mimicry is clearest in the case of transnational entities such as Oxfam or Greenpeace, who use business-style strategy to promote "efficient" resource allocation and organizational survival under competitive conditions (Wapner 1996; Jordan and Maloney 1997).

This diverse literature shows how the need for money can distort missions. It corrects the view of nonprofits as "anti-commercial" and "anti-establishment." Indeed, I confirm below this influence of money. Yet, one must not overstate the importance of economics. My aim in part is thus to show how a concern for reputation may prompt "irrational" choices. There is therefore an ambiguous quality to the financial strategizing of NGOs.

Funding Autonomous Action

There is often an image conveyed in the literature of NGOs trapped in "unseemly" pursuit of funding (e.g., Smillie 1995; Sogge 1996a). This image is important for two reasons. It is revealing because of the implicit criticism involved—a view that NGOs ought not to behave so because perceptions of blatant self-interest might clash with other perceptions of moral and altruistic intent. Yet the image is also significant for what it does not say. It fails to consider the creative way that NGOs seek funds so as to safeguard a valued reputation. In other words, this image does not acknowledge that the quest for moral capital and the search for cash are often intertwined.

Much strategizing reflects the promotion of autonomy through shrewd fund-raising and spending. I consider below this campaign as well as limitations associated with donor dependency. It is worth considering first why financial autonomy might be attractive in relation to moral capital. To possess autonomy is to suggest that an organization is not beholden to others in a way that compromises an organization's mission. This is important since NGOs are often identified as, and valued for, being relatively free to promote change (Korten 1990; Fisher 1998). Autonomy can also be of general benefit insofar as partners respect an organization that "stands on its own two feet." This respect may enable the accumulation of moral capital as relative autonomy boosts the credibility of an organization—even if this is but one element in a larger process (Fowler 2000).

How do NGOs seek autonomy through the raising and the spending of money? There is "no single strategy or recipe" but rather "combinations of multiple strategies" depending on specific circumstances (De Rosas-Ignacio 1997: 3). Three aspects are to be noted. First, there is promotion of *a range of income streams* to reduce dependency. Second, there is promotion of *"efficient" expenditure procedures* to maximize benefit from funding. Finally, funding is sought from various donors—especially those who permit flexible spending—to *reduce dependency on any one donor*. All of which promotes what Fowler (2000: xii) calls "insightful agility"—the pursuit of financial sustainability when NGOs "continually adapt and adjust in a purposeful, not random, way." As we shall see, there is often a mixed record here. Of particular interest is how the pursuit of goals relates to the quest for moral capital, resulting sometimes in difficult tradeoffs.

Philippine NGOs have notably promoted the goal of diversified income streams by experimenting with initiatives involving the general public. True, compared with the experience of NGOs in the West, they have scarcely made headway—initiatives add up to only a tiny proportion of total income. Yet consideration of "peripheral" endeavors such as membership drives or sponsored films suggests that there may be dividends in terms of moral capital. The experience of the Haribon Foundation—a national pioneer at public fundraising—illustrates the possible strengths and weaknesses of strategies involving public appeals. Central here is a scheme whereby individuals and groups (usually businesses) become voting members of the organization. Membership numbers have certainly fluctuated: a few dozen in the 1970s, over 800 at the end of the 1980s, but only 127 in 1997 (Mangulatron 1996; Holopainen 1997). Membership has been largely student based, with "chapters" run by school and university students. By itself, membership did not increase income at the national office, since student dues were low and retained by individual chapters for local use.

Yet membership has a far wider significance here. Urban middle-class members are targeted to boost income through promotional products and events. Thus, the member's magazine features information on sponsored films and Haribon products. Members are also the "foot soldiers" of protest, with students a vocal presence at demonstrations. The impact of this free labor is not to be gainsaid. Finally, the membership is a means to spread the Haribon message. Members tell families and friends about the NGO (Holopainen 1997). The activities of chapters provide another avenue for information dissemination and reputation building. For example, students at the University of the Philippines Los Baños sponsored a film in 1989, the proceeds from which sustained a high-profile campus trash project. The NGO used its members to re-create the "Katipunan effect"—a process in which more and more people would learn about and join the organization (Haribon Foundation 1990a: 4).[1]

The Katipunan effect is also sought through special events such as rock concerts or films. The advance screening of Hollywood films with an environmental theme is a particular specialty. Benefit events have included *The Mosquito Coast, Fern Gully—The Last Rainforest,* and *Gorillas in the Mist.* The usual practice is to dedicate income from ticket sales and souvenir programs to specific activities. For example, money earned from *Fern Gully* went to the Mount Isarog project, while proceeds from *Gorillas in the Mist* went to Tanggol-Kalikasan. These events were an opportunity to publicize work and to promote environmental knowledge. Thus, in promoting *Gorillas in the Mist,* the work of Tanggol-Kalikasan as an advocate of environmental claims was highlighted next to an appeal for wildlife conservation. Here, then, is a modest income as well as useful publicity.[2]

The sale of promotional products has been motivated by an attempt to boost both income and name recognition. The Haribon logo is plastered over t-shirts, pens, stationary, postcards, gold-plated pins, stickers, backpacks, sleeping bags, hip packs, fanny packs, and "yuppie bags." The rare Philippine eagle (after which Haribon is named) and the Tarsier (an endangered small primate) have been pressed into service to sell posters, postcards, and letter openers. For the affluent, there is a framed Philippine eagle print that, it is suggested, would "make very impressive wall décor and conversational piece for visitors" (Haribon Foundation 1988: 18–19). Then there are the scientific books and the NGO's *Enviroscope* magazine for those of a scientific bent.

True, such goods have produced a miniscule income. There was difficulty with flogging the expensive eagle prints, for example. It has been nonetheless an additional means of boosting the name and cause of the NGO. As the likes of Nike and Calvin Klein demonstrate, the wearing of corporate logos on clothing, shoes, or bags is important both as a source of free advertising and as a testament to the socially embedded nature of a firm (Klein 2000). In a similar albeit more modest way, Haribon benefits from the circulation of its "brand name" in Philippine society as people wear its products. Still, public outreach of this kind has limits. It targets a small if affluent urban middle class. This focus may be financially astute but does little to promote either name or message to most Filipinos. Indeed, it may smack of elitism. As the Haribon's Christi Nozawa (1996) remarked, fund-raising tactics were "not the same for the middle classes as they were for others in society." And yet, there was little sense of how those not in the middle classes fit into the picture. Further, there is no tradition of secular donation even among the affluent, since money is largely channeled through churches (some of which gets to NGOs)(Fabros 1988; Alegre 1996b; Clarke 1998).

The Haribon Foundation nonetheless remained committed to

public initiatives that promoted itself and its cause. The environment is indeed a salable commodity with the middle classes—a fact used to promote financial sustainability (Honasan 1996; Connell 1999). One leader explained that the goal was to boost membership so that dues would "support operations," thereby "avoiding external dependency," since reliance on donors hindered "political advocacy" (Tongson 1996). This was wishful thinking, in that only one employee handled membership and the total income from fund-raising and membership fees in 1996 amounted to $2,363—compared, for instance, with other income dedicated to administrative overheads in that year of $37,654 (Haribon Foundation 1997a: appendix d).

If Haribon spends "disproportionate" time on such fund-raising given the paltry financial return, it is because there are other benefits. The income is not tied to predetermined expenditure plans, as with donor funding, thereby permitting room for discretionary spending. There can be benefits too in the form of free labor and publicity. However, this is not the only way in which to promote room for maneuver. Another way is shrewd spending procedures that maximize the social and economic return on income. And yet, "insightful agility" here can lead to ambiguous tradeoffs in the quest for moral capital.

The record of the PAFID is illustrative. Unlike Haribon, it does not devote much effort to public fund-raising. Yet it has generated modest wriggle room for itself through careful attention to expenditure. One way is through what Rice (1996) dubs "creative financing." Here, the NGO takes advantage of the fact that large sums of money flow through its bank account as donors provide funds for onward disbursement to local communities or NGO employees. In that there is an inevitable delay between receipt and disbursement of funds, a small income is earned through interest accumulated. The PAFID began this process in the mid-1980s when it received its first large grants. In 1993, at a time when income was at an all-time high, such interest amounted to several thousand U.S. dollars (PAFID 1994a). This process is neither illegal nor injurious to projects. Indeed, donors often approve of it. As the Canadian International Development Agency's Jun Braza (1997) explained, "creative accounting" by NGOs working with his agency permitted them "to retain savings for future activities." However minimal the sum in the end, this process is another means to create some financial space.

There are also "creative labor practices." NGOs usually depend on the retention of motivated albeit poorly paid staff. The gap between what is paid and what is offered for comparable work elsewhere varies. A key factor is the type of work undertaken. Here, the gap may be smallest for community workers and largest for those with accounting or legal skills. One administrator remarked that the salary of a friend doing comparable work in the business sector was "approximately double" her own pay, while an employee involved in legal work estimated that her salary was one-third of counterparts working for private law firms.[3] Whatever the "opportunity cost," though, NGO work is usually predicated on personal financial sacrifice. True, there can be other benefits. Comments such as "I always wanted to do environment work" or "I believe in Haribon" or "I love the PAFID" were common, although such statements required guarded usage.[4] Further, while individuals certainly defect to business or government, the NGO sector is undoubtedly underpinned by personal financial sacrifice and cause-related devotion (Brett 1993; Fowler 1997).

NGOs can thrive even in financially uncertain times. In fact, there are decided advantages when employees are subject to "super-exploitation." There are *cost savings* on the labor bill—by far the main expenditure—as individuals put in long hours for relatively poor pay. Experience at the PAFID is probably typical—at least for mid-to-large Philippine NGOs. In the mid-1990s, the standard working week at the NGO was between sixty and seventy hours without provision for overtime pay. There was, though, three weeks paid holiday on top of Christmas as well as an allowance of fifteen days for paid sick leave. This was an important and necessary provision, given grueling and often hazardous employment conditions. The *flexible* nature of employment further enables organizations to adapt quickly to new circumstances. For example, they can shed jobs at the end of a project when new funds are not obtained. At the PAFID, hiring is "coterminous with projects" (Tindungan 1996). Thus, when it was decided to pullout of a project funded by the Asian Development Bank (discussed below), nearly half of the staff was laid off. In the absence of permanent contracts and unions, there was little to do except look elsewhere for employment—often with the help of the PAFID's management.

In this way, NGOs do more with less. The moral implications are not straightforward, though. An NGO may benefit from perceptions that employees make sacrifices to further its mission. Thus such sacrifice may be viewed as "living proof" that the organization is there primarily to help others. Indeed, donors regularly extol the "dedication," "sense of commitment," "selflessness," and "conviction" of NGO staff who accept low salaries for hard work. Local leaders comment favorably on NGO employees showing "dedication" and commitment to their communities even under difficult circumstances. Residents tend not to respect those that are "here one day and gone the next."[5]

Still, there are moral ambiguities involved here. While organizations may benefit as discussed, employees bear the burden of the sacrifice. This is especially hard on those with families to support. As one person anonymously observed, the salary at his NGO, although better than at some organizations, was "still inadequate," necessitating freelance work to fill the gaps. Fortunately, many organizations allow employees to undertake consulting work on the side. At the PAFID, for instance, they are encouraged to develop "alternative

Figure 5.1
NGO Strategic Options in Resource Mobilization

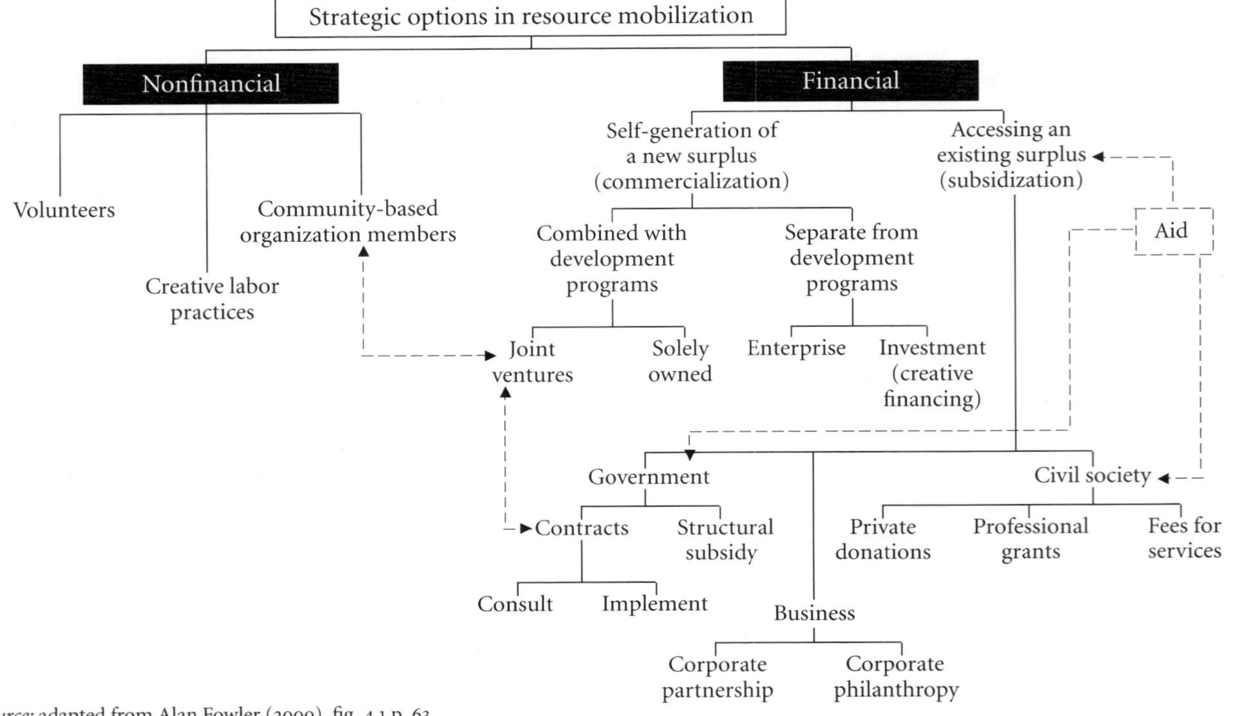

Source: adapted from Alan Fowler (2000), fig. 4.1,p. 63.

livelihoods" linked to research for donors (Tindungan 1996). This NGO also ran a staff credit cooperative to enable employees to borrow in emergencies or for house purchases. Similarly, Haribon allows staff to conduct legal or environmental consulting work to top-up salaries. Yet the demands of regular employment may clash with the need to supplement income. At the PAFID, for instance, few employees avail themselves of the alternative livelihood scheme because they have little time to do so. NGOs such as the PAFID or Haribon acknowledge financial difficulties faced by employees. Yet little is done to lighten workloads to facilitate the pursuit of extra income. Such "efficient" expenditure, then, raises troubling ethical questions about staff treatment even though it enhances the financial welfare of the organization itself.

The strategies discussed thus far share a concern: the promotion of organizational financial autonomy. If complete self-financing is a pipe dream, there are nonetheless a range of options that NGOs can adopt to maximize their room for maneuver. Before turning to donor-oriented strategies, let me briefly assess the range of available options relative to their strategic importance and desirability (see figure 5.1). Figure 5.1 prompts three observations. First, NGOs use nonfinancial means to promote autonomy. In reality, though, there are social, economic, and ethical implications of doing more with less. Second, they seek to generate income through their own funds for discretionary spending (although our example above was both modest and parasitic of donors). Finally, NGOs enlist support from diverse donors, notably to dilute dependency on state and business-linked sources by boosting income from civil society (e.g., NGO-managed funding mechanisms, cooperatives, or public memberships).

How do these options relate to the interlinked pursuit of money and moral capital? The quest for moral capital is likely to condition how NGOs raise funds. Thus the search for cash is not simply a matter of maximizing income, since financial sustainability involves a broader endeavor of relating potential funding to anticipated implications for reputation. What effect does acquiring funds from one source have on the ability to accumulate moral capital with others? Does the choice build financial autonomy? Are strings attached and with what ethical implications? The point here is *not* to suggest that NGOs always privilege moral concerns (and perceptions thereto) in decision-making. Rather it is to underscore tradeoffs between economic necessity, financial probity, and moral purpose that bedevil thinking in the NGO sector (Bennett and Gibbs 1996; Hulme and Edwards 1997; Fowler 2000).

To determine how funding options relate to moral capital it is necessary to consider criteria of relative desirability and importance. The relative *desirability* of an option suggests NGO preference. Given ideal conditions, which option is best suited to the goals of autonomy and reputation building? In contrast, the relative *importance* of an option relates to "real-world" conditions. Which option is most critical to NGO operations? Table 5.1 provides an explanation of possible tradeoffs. It represents a *plausible* explanation and therefore does not purport to represent an empirical validation. Its utility resides in the logic that it suggests underpins NGO strategizing. Table 5.1 suggests that options are rarely valued according to the criteria in an identical fashion. An option may be highly desirable but of little importance. For Haribon, as noted, public donations via membership, special events, or promotional products amounted to a tiny fraction of income. Yet there were perceived benefits: discretionary income as well as an opportunity to boost public name and cause recognition. Public fund-raising can thus be desirable even when employees know that it will never be a major money earner. Its social significance outstrips immediate economic value, leading to seemingly "irrational" behavior.

In contrast, an option of moderate to high importance may be

TABLE 5.1 The Relative Desirability and Importance of NGO Funding Options

Option	*Desirability*	*Importance*
Public fund-raising	High: discretionary (+); publicity (+)	Low
ODA	Medium: amount (+); terms (−)	High
NGO managed	High: amount (+); NGO control (+)	Medium to high
INGO	High: NGO control (+)	Low
PPF	Low: adverse publicity (−); ethics (−)	Low to moderate
IPF	High: amount (+); flexibility (+)	High
Philippine govt.	Medium: amount (+); terms (−)	Medium

Note: ODA = overseas development assistance; INGO = international NGO; PPF = Philippine private foundation; IPF = international private foundation.

less desirable because there are social costs associated with acceptance. These may include reduced autonomy, a distorted mission, or a tarnished reputation. One example is where money is drawn from businesses or allied "charitable" foundations but NGOs thereby become linked to "disreputable" activities. Even donor funding may be viewed unfavorably because of the politicized identity of the donor or unattractive conditions. The USAID—part of the U.S. government—has been viewed thus by some NGOs. Finally, funding controlled by the NGO sector itself is likely to be desirable and increasingly important. NGO-managed funding mechanisms enable organizations to decide on priorities among themselves. Yet, mechanisms such as the Foundation for the Philippine Environment and the Foundation for a Sustainable Society can intensify competition between NGOs.

Most NGOs remain nonetheless dependent on donors. That they are largely dependent on *foreign* donors is an added concern. As one 1997 headline of *Philippine NGO Memo* proclaimed, financial sustainability is "a declaration of independence" from external funds. Critics use such dependency to attack NGOs. In one newspaper, for instance, they were seen as "society's new 'rent-seekers' and parasites" inasmuch as they were "too dependent on foreign funding and subsidies" (Jimenez 1993: 18). Elsewhere, a "full accounting" was demanded of funds received by NGOs that "target direct foreign assistance without explicit accountability to the Philippine Government"—with foreign money even "subsidizing the communist insurgency" (Dolor et al. 1994a: 7).

Donor dependency does not mean all room for maneuver is lost. Indeed, NGOs can win funds and boost moral capital by careful strategizing. A key factor here is donor heterogeneity: foreign aid agencies and embassies, international financial institutions, private foreign organizations (such as international NGOs or church groups), foreign and domestic business, national and local government, or Philippine church organizations. There are diverse priorities here—something upon which an astute NGO may capitalize (Smith 1990).

There is hence incentive to develop a diversified donor portfolio. This strategy limits exposure to any one donor, thereby reducing the likelihood that a mission is distorted by donor priorities. It also widens the pool of partners, thus increasing the likelihood that an organization's reputation will become known. Doing the rounds of the donor community is the norm as NGO leaders seek to keep funding rolling in (Alegre 1997). In this regard, the quest for moral capital and the need for donor income may reinforce each other while not being identical. I assess below the moral ambiguities involved, but here I focus on strategic aspects to the search for diversified donor support.

In the case of Haribon, the pursuit of environmental, community-organizing, and paralegal interests has led it to seek funding from a range of small, medium, and large donors. Table 5.2 provides a summary for the first half of the 1990s. There is wide-ranging support for Haribon, which is reflected in the diversity of donors with whom it works: official overseas development assistant (ODA) providers, embassies, international NGOs, the Philippine government, NGO-managed funding bodies, United Nations organizations, international research institutions and funding organizations, and international private foundations. This diversity is also often reflected at the level of individual projects. A key grant and donor relationship is supplemented with additional grants from others. In the case of the project to run a course on coastal management, for example, the main source was the Rockefeller Brothers Fund, with additional funds provided by the United Kingdom embassy.

There is a strong international aspect to this profile. If the sector as a whole is reliant on foreign donors, this appears doubly so in environmental management, where NGOs benefited from international environmentalism. As the country's pioneering environmental NGO, it is perhaps not surprising that this organization is especially dependent on foreigners. Still, it is vulnerable to domestic criticism of foreign rent seeking, as noted above. Such criticism can help to erode the stock of moral capital that Haribon has accumulated with domestic actors inasmuch as the charge of foreign dependency chimes with wider views on outside intervention (Clarke 1998). As one former DENR secretary remarked, NGOs such as Haribon may be "too dependent on foreign NGOs and donors" in what amounts to "a new form of neo-colonialism" (Factoran 1997). And yet, Haribon is also vulnerable to shifting donor preferences. It has thus been criticized for foreign dependency even as it has suffered from donor reassessments of Philippine funding—thereby providing a particularly fraught context of fluctuating moral capital and donor assistance as a result of two distinct processes.

Yet there are advantages too, since Haribon's funding has been less dependent than usual on shifting political and economic networks *within* the Philippines. To some extent, this provides the NGO with breathing room, especially as it interacts with state partners. This can lead, in turn, to a greater ability and willingness to

TABLE 5.2 Haribon Foundation Funding (Selected Projects)

Donor	*Type*	*Purpose*	*Amount*	*Comment*
Mt. Isarog National Park				
WWF/DENR	INGO/Govt.	CO	P2,042,900	Debt-for-nature
FPE	NGO-managed	CO/legal	P1,657,242	
Finnida	ODA	livelihood	—	
PACAP	ODA	livelihood	P750,000	
Marine Conservation: Zambales				
Netherlands Embassy	Embassy	CO	—	
DENR	Govt.	training	P750,000	
CBCRM: Pangasinan				
FAO	UN	CO	P1,103,000	Bolinao site
IDRC	IRI	training	P3,800,000	
CBCRM: Batangas				
WWF/DENR	INGO/Govt.	CO	P3,812,532	Debt-for-nature
Conservation, Philippine Birds				
Bird Life International	INGO	research	P208,000	
Wild Bird Society of Japan	INGO	research	P208,000	
U.K. Embassy	Embassy	research	P570,000	
Sea Horse Conservation				
Darwin Initiative	IRI	research	P320,000	
Oxford University (U.K.)	IRI	research	P270,219	
Aquarium Fishing				
IDRC	IRI	training		
WWF/DENR	INGO/Govt.	training	P2,443,825	Debt-for-nature
Ocean Voice/CIDA	INGO/ODA	training	P1,768,000	
Course, Coastal Management				
Rockefeller Brothers	IPF	training	P5,200,000	
U.K. Embassy	Embassy	training	P1,500,000	
Coastal Environment Program				
DENR	Govt.	training	P400,000	
Capacity-building course				
FPE	NGO-managed	training	P411,300	
Tanggol-Kalikasan				
Asia Foundation	IPF	legal/training	P5,800,845	
FPE	NGO-managed	legal/training	—	
Misereor	IPF	legal/training	P169,280	
W. Alton Jones	IPF	legal/training	—	
Friedrich Ebert Stiftung	IPF	legal/training	—	
Endefense Program				
FPE	NGO-managed	litigation	—	

TABLE 5.2 *Continued*

Donor	*Type*	*Purpose*	*Amount*	*Comment*
Science & Litigation				
Asia Foundation	IPF	training	—	
Renewable Energy Study				
Carl Driesburg Gesselschaft	IPF	research	P480,000	
Core & Miscellaneous Support				
MacArthur Foundation	IPF	core/project	P16,828,192	1991–1996
ILO	UN	training	P302,715	

Source: Haribon Foundation *1996 Year End Report*, Annexes A & L; Haribon Foundation, "Project Profile" (Manila: Haribon Foundation mimeo, 1997).

Note:
The table covers selected projects only; most donors/funds included here.
Funding by main purpose only.
Conversion rate used: $1 = P26.
CO = community organizing; INGO = international NGO; ODA = overseas development assistance; UN = United Nations; IRI = international research institute; IPF = international private foundation; CBCRM = community-based coastal resource management.

criticize powerful political and economic actors than would be the case if the NGO relied on them for funding support (Broad with Cavanagh 1993). Here, then, is a greater chance to accumulate moral capital through "critical engagement," provided that it is tailored to the policy or issue in question (Aldaba 1997; see chapter 4). True, foreign donors can be chary of "political NGOs." As one European donor forcefully observed, these NGOs adhered to "outmoded ideological dogma" and were "insulated from society"—and hence unworthy of donor support (Teunissen 1997). In the case of the MacArthur Foundation—Haribon's main backer in the 1990s—there was indeed explicit prohibition against overt political advocacy by partners. Yet both parties interpreted this prohibition flexibly. One Haribon leader described MacArthur as "quite generous," with "no overt strings attached" to funding (Nozawa 1996). True, it is only politic to praise donors. However, and as documented at various points in this book, MacArthur Foundation funding coincided with some intense environmental campaign work by Haribon of a distinctly political nature. Here again, it is getting the balance right in mixing criticism and cooperation that often holds the key to an NGO's moral capital.

Haribon has also eased the task of developing diversified donor support by going global. Here, there is a wide array of donors. Thus funding is obtained from overseas development assistance (ODA) providers, embassies, research centers, United Nations organizations, private foundations, and NGOs. Further, funding is obtained from donors based in various countries: Japan, Finland, the Netherlands, the United States, the United Kingdom, Germany, Australia, and Canada. This multifaceted profile enhances Haribon's room for maneuver vis-à-vis the donor community (including international NGOs such as the World Wildlife Fund (WWF) and Conservation International). That funding is obtained from private and public sources in North America, Europe, and Australasia is especially significant insofar as it ensures that one set of foreign interests never enjoys overwhelming influence. Further, this differentiation in overseas support also enables the NGO to pursue moral capital with a wide pool of partners. There are two aspects to note here. On the one hand, donors often check with each other before supporting funding applications—giving a pioneering and well-known NGO such as Haribon a potential head start over lesser-known competitors. On the other hand, this is still a fragmented decision-making process with diverse donor interests eliminating the prospect of a monolithic donor position. There is some scope, in other words, for shopping around in the search for cash so as to minimize the risks of seeing missions distorted due to excessive donor dependency. The latter, for instance, was a recurrent worry of the board of trustees (e.g., Haribon board minutes 1991). Either way, the diversified overseas donor portfolio held attraction for Haribon precisely because it suggested an efficacious means to accumulate moral capital.

Such talk of diversified donor portfolios must not be exaggerated though. Both Haribon and the PAFID have been somewhat reliant on key donors for critical income streams at various stages despite having ostensibly diverse donor portfolios. Here, the specific *quality* of the funding is at stake. The MacArthur Foundation provided 12 percent of Haribon income in 1996, while the German Catholic aid agency Misereor provided over 24 percent of PAFID income in 1990 (Haribon Foundation 1997a: annex l; PAFID 1991: n.p.). Both grants represented a long-term commitment that provided core funding—"highly critical resources . . . [not] easily replaced" (Fowler 2000: 62).

However, such dependency need not mean that autonomy is reduced nor that NGO credibility is jeopardized. Everything depends on the nature of the dependency in question. Thus, if an organization is dependent on a "blue-chip" donor, then autonomy and credibility with an array of partners may even be enhanced. Such a donor is valuable for several reasons. Funding is usually given for core operations over three or more years, thereby enhancing financial certainty. There are usually few strings attached, thus permitting some discretion in spending. Indeed, the donor can be quite tolerant when an NGO deviates from agreed practices, provided changes are justified. Not surprisingly, blue-chip donors are simultaneously highly desirable and important to NGOs (see IPFs in Table 5.1). They are mainly international private foundations (IPFs), because others rarely have the funds (NGOs) or flexibility (ODA providers) to provide such support.[6]

Both Haribon and the PAFID were supported thus in the early 1990s. The MacArthur Foundation sustained Haribon by covering nearly half of administrative costs (the rest was claimed through projects). I noted above how this relationship featured flexibility over political advocacy. MacArthur support also provided financial breathing space while employees reassessed the identity of the organization (nearly abandoning the Haribon name in the process) (Nozawa 1996; Tongson 1996). Misereor played a similar role for the PAFID at this time. This Catholic funding agency based in Germany has been a major supporter of work on behalf of indigenous people, and its grants funded community organizing, technical support, staff training, workshops, political lobbying, and administrative reorganization. The PAFID certainly needed to meet targets in terms of land tenure applications, communities organized, and so on. Yet targets were set in consultation with NGO employees, thereby permitting "leeway" in expenditure—for example, in terms of which communities were organized (PAFID 1995c: 2). Such support was invaluable, as it meant employees could pursue a well-rounded campaign on behalf of indigenous partners (De Vera 1996; Rice 1996).

Blue-chip funding was seen to avoid the bureaucratic hurdles surrounding much ODA. At USAID, for example, "stringent reporting requirements" and the rule that NGOs provide 25 percent matching funds put many off (Magno 1997). Even USAID (1997: 9, 11) noted that "grantees often resent" arrangements that are "burdensome." The PAFID experience here is illustrative. The NGO received a one-year grant in the mid-1980s, but USAID encouraged them to apply for a further three-year grant. The PAFID was put off the idea, though, on two grounds. First, acceptance might place employees in jeopardy since the New People's Army was hostile to U.S. projects. Then PAFID leader Delbert Rice (1997) sought a solution by asking USAID not to publicize the grant; however, "they would not make such a promise" because they "needed the PR [public relations] impact." Here we see a situation in which these two actors had conflicting needs concerning moral capital, pointing to a tension in the entire process. Second, PAFID staff "had problems" with the paperwork: "We had to carry a magnifying glass with us in order to read the fine print in the agreements, some of which were hardly intelligible even when we could read the words" (Rice 1997). The PAFID thus declined the follow-on agreement. Yet there was to be a further sting in the tail. Two years after the end of the first agreement, a USAID "re-audit" rejected selected PAFID expenses not "properly audited at the time the project was closed." It was thus forced to refund money to the agency (Rice 1997).

Not surprisingly, the reputation of USAID among many NGOs was poor—a perception exacerbated no doubt by general anti-American feelings linked to the colonial and postcolonial experience (Wurfel 1988; Go and Foster 2003). Indeed, NGO maneuverability was further curtailed, since contracts usually require "specified deliverables." Yet, if these sorts of rigid bureaucratic procedures disenchant NGOs, there is only so much blue-chip money to go around. The entangled quest for money and moral capital often lead NGOs into relationships that privately they deplore.

All of this occurs against the backdrop of the worldwide spread of neoliberal ideas about governance. The so-called "counter revolution" in political and economic thinking that has gained momentum since the 1970s has permeated most aspects of contemporary life and has centered on "the reassertion of the private over the public interest" (Taylor et al. 2002: 11; see also Toye 1993; Dicken 2003). As a "new managerialism" has taken root, NGOs have played a pivotal role in the process whereby states have been "downsized" and/or "reformatted" to better reflect the concerns of global capital (Taylor et al. 2002). Two aspects may be briefly noted here. On the one hand, NGOs have facilitated this process through involvement in a "contracting culture" that substitutes the purported virtues of "low-cost," "flexibility," and "participation" associated with the nonprofit sector for the alleged vices of "inefficient" and "bureaucratic" management associated with government (Robinson 1997). On the other hand, NGOs have themselves been "disciplined" in the process through requirements that they become more like private firms through notions of "efficiency," "accountability," and "professionalism" (Edwards and Hulme 1995; Anheier and Salamon 1998; Fowler 2000; Smillie and Hailey 2001). Philippine NGOs have not been exempt from this process. They have certainly benefited from the upsurge in funds available to the NGO sector as a result of neoliberalism (Clarke 1998). And yet they have faced donor-led efforts to impose business-style notions (as suggested above) on their sector. The general reaction has been to decry the imposition of "alien" management forms that, as the CODE-NGO's Karina Constantino-David (1997) puts it, "destroys alternative-ness." How, though, does this trend relate to NGO efforts to pursue funding and moral capital?

One view is that businesslike practice forces NGOs to be efficient users of funds in a process that helps beneficiaries. New budgetary requirements instill "discipline" and "evolve from being a burden to a tool in project management" (USAID 1997: 9). Such requirements include regular financial reports, standard financial procedures, and annual audits by independent accountants. Neither Haribon nor the PAFID has been immune. Thus one objective of Misereor support was to "improve the organizational sustainability and efficiency" of the PAFID, involving a "human resources development program" plus upgrading of "administrative, operational, and information systems" (PAFID 1995c: 3, 23–24). Work undertaken in 1995, for example, included: completion of an orientation handbook; color-coding of files; the introduction of computerized bookkeeping; and two financial audits. Haribon responded similarly to the wishes of MacArthur and other donors, producing an employee manual as well as a financial policy manual setting out accounting and financial control systems—a document that was also a benchmark for auditors (Haribon Foundation 1997a).

This process did not occur without protest. Constantino-David (1998: 46) argued that "the underlying assumptions of these practices—hierarchy, profit, suspicion—go against the grain of voluntarism" even as "greater percentages of limited budgets are allocated to financial and administrative support staff [while] direct service personnel are swamped with bureaucratic requirements that take up precious time." Resistance was not therefore unusual. Some Haribon staff, for example, fought a new time-keeping system designed to better monitor employees. A "daily time record" in which staff logged in and out was especially resented—and was soon dropped in favor of a "monthly time planner and organizer" (anonymous interviews). Still, morale of employees suffered under new notions of labor discipline.

The impact of these changes on moral capital is not clear-cut. There is usually a more favorable impression of NGOs among donors—hence, that they are "more professional" and "more effective" than in the past and "easier for donors to communicate with" (Chua 1997; Racelis 1997). Another added that NGOs had learned that "to be effective, you must also be efficient" (Braza 1997). In addition, when NGOs are publicly seen to be thus accountable, they can parry somewhat the accusations of opponents that they lack accountability. As one NGO leader pointed out in rejecting the idea of official monitoring, there was no need for it since "our funders are quite strict about monitoring and evaluation" (Dolor et al. 1994b: 5). And yet, there is no demonstration that such "efficiency gains" lead to better service. Indeed, to the extent that resources are simply diverted from frontline service to management, field staff face a harder than usual time in their work (Constantino-David 1998). As employees devote more energy to paperwork than to project work, the task of accumulating moral capital with local partners particularly becomes more difficult. Certainly NGO employees complain of the distractions of paperwork (anonymous interviews).

The "efficient" NGO is thus not necessarily best suited to the multifaceted quest for moral capital. Yet dilemmas associated with neoliberalism must be situated in the context of efforts by NGOs like the PAFID and Haribon to boost autonomy and moral capital through shrewd fund-raising and spending. Not that strategizing gets them off the hook. These strategies can be only partly successful and even raise their own moral ambiguities. What comes across in the discussion thus far is that NGOs face a range of choices when thinking strategically about money. Just as moral capital may be accumulated with diverse actors through the choices that NGOs make in the pursuit of funds, those choices themselves often reflect the imperatives of being seen to be a moral and altruistic actor. This point—which is a world away from the view that says NGOs will do anything for cash—is perhaps most evident in those rare yet revealing cases in which NGOs turn down funding opportunities.

Choosy Beggars

That an NGO would decline funding even if hardship would result makes sense when its role as a moral entrepreneur is acknowledged. "Irrational" behavior is comprehensible when the search for cash is seen as linked to a multifaceted quest for moral capital. While not-for-profit status condemns NGOs to a life of fund-raising, funds are rarely accepted without regard for the circumstances. Concern about reputation influences how cash is acquired. A dramatic demonstration is when an organization withdraws from financial commitments. This option is not chosen lightly, since there can be substantial penalties as relations with state agencies and donors turn sour. A reputation for unreliability may be hard to shake. So why risk such an outcome? Examination of one PAFID experience is helpful here because it shows how an NGO may reject financing when it is perceived to adversely affect pursuit of a mission and moral capital. The case also shows the strain on staff that this entails.

The Low-income Upland Communities Project (LIUCP) ran for eight years, beginning in 1990 with $32 million provided by the Asian Development Bank (ADB) and $8 million from the Philippine government. It promoted upland conservation and economic security for Mangyan people living on Mindoro. Ultimate control rested with the Department of Environment and Natural Resources (DENR) as well as the provincial governments of Occidental and Oriental Mindoro. And yet, implementation was by six NGOs, among which featured the PAFID. The project itself involved community organizing and cooperative development; resource access and resource management; agroforestry, reforestation, and livelihood; infrastructure and social services; and project management and institution building (De Vera 1994; Marco 1994; PAFID board minutes 1994).

The LIUCP started amidst high hopes. For the ADB and the DENR, it was an ideal opportunity to publicly demonstrate their support for a participatory style of intervention. As the ADB's Tom Waltz (1997) claimed, this project was an example of the "bottom up approach" that the ADB "ardently embraced" in the 1990s, requiring input from NGOs. For the Mangyan, it was a potential means to reverse land-grabs by local bosses (Rood 1998). For NGOs, it seemed a chance to rectify unjust treatment of indigenous people. As the PAFID's Willy Tolentino observed, the project offered "an opportunity to help uplift the socio-economic condition of the Mangyans while addressing ecological problems and issues"—even as his organization would "assist the people in the preparation of their own development plans" (cited in LIUCP 1993: 4–5; Tolentino 1997). However, the LIUCP was soon beset by conflict, as different priorities emerged. After three years of acrimonious interaction, the NGO coalition (the United NGOs of Mindoro or UNOM) withdrew from the LIUCP in December 1993, much to the anger of the DENR and the ADB. UNOM developed an alternative framework through the Kapulungan Para sa Lupaing Ninuno (KPLN)—a Mindoro NGO–people's organization (PO) forum on ancestral domain.

What to make of this episode? At one level, it is an instance of conflicting priorities. There were various misunderstandings surrounding slow processing of land tenure documents as well as delays in canceling pasture leases. An irritant was the delayed disbursement of funds for livelihood schemes. On a number of occasions, the PAFID had to pay people's organizations out of its own funds to make up for delays at the DENR. Rice (1996) recalled that at one stage the PAFID was owed one million pesos (U.S.$40,000). There was also bureaucratic foot-dragging by the government on "low" priorities such as livelihood support or land tenure. The noninvolvement of NGOs and POs in policy formulation led to unrealistic expectations being placed on them. This is clear from remarks by the ADB's Tom Waltz (1997), who suggested that difficulties reflected official expectations that "Mangyan productivity would be higher" and that "NGOs [were] not doing their job effectively." However, for Tolentino and other PAFID staff, the problem was that government and the ADB did not allot enough time for community organizing. This slipup reflected a "narrow" outcome-oriented rather than process-oriented framework that was monitored by ineffectual consultants. As one PAFID officer remarked, the main problem was that, by pushing reforestation over livelihood, the DENR gave the impression that "forests were more important than people" (Vargas 1996). The LIUCP thus "failed to overcome the mutual distrust which prevailed . . . rendering supposed tripartite relations ineffective" (Marco 1994: iv).

The LIUCP speaks volumes about the priorities of NGOs when confronted with a seeming no-win situation. On the one hand, the LIUCP was an opportunity to "scale up," in that NGOs could expand operations and hire staff—all in aid of helping the Mangyan to win ancestral rights (Constantino-David 1992). On the other hand, the LIUCP required NGOs to fit into management structures that Mangyan found to be a big problem. As the debilitating effects of those structures became apparent, the PAFID and others were caught in the middle between angry communities and complacent bureaucrats. For example, one community blamed the PAFID when delays in payments by the DENR disrupted planting activities (Vargas 1996). The NGOs thus faced a dilemma: fight to turn the project around while risking criticism from communities or withdraw altogether, thereby ending such criticism at the price of soured relations with the government and the ADB.

At the PAFID, political calculations mixed with moral reflection in the decision to pull out. It was, as PAFID leader Shirley Libre suggested, "an ethical issue" that went to the heart of what the NGO was about (cited by Rice 1996). True, it was wrong to break a contract. But it was even worse to betray indigenous partners. Defending the latter was the source of the PAFID's ability to effect change—its "gold mine" (PAFID 1994c: 12). The pullout was therefore designed to safeguard that "gold mine" by demonstrating solidarity with local communities, even if a penalty would ensue.

The decision nonetheless provoked debate. There were discussions among Mindoro staff as well as with Manila. The PAFID board met twice on this issue. In July 1993 it advised on the stance that the PAFID should take at an upcoming meeting of concerned NGOs to agree on the NGO coalition position. In November the board prepared a letter of termination in light of the clear wish of PAFID field staff to escape from a deteriorating situation. Still, there was division at the top. A key division was between Libre and Rice. In Libre's view, involvement had to end because of DENR mismanagement. Rice (1996) believed that Libre and other staff "overreacted," allowing things to get "out of hand." While acknowledging deficiencies at DENR, Rice believed that differences could have been resolved through diplomacy. Many staff argued nonetheless for a pullout. Thus "objectives were not achieved," in part due to "inadequate consultation" with NGOs and people's organizations even while there was "no political will to respond" to these complaints; the organization was thus "wasting time" with the LIUCP (Vargas 1996). In light of these views, Rice sought to ensure instead that the letter of termination was written tactfully to minimize damage to the PAFID's reputation at the DENR. Here Rice (1996) insisted on major revisions to the draft letter prepared by Libre so that it was "toned down."

The resulting letter to DENR Secretary Angel Alcala of November 25, 1993, was a model of the sort of critical engagement discussed in chapter 4—applied here to the high-profile rejection of funding. It merits careful consideration (Libre 1993). The six-page letter was written in an upbeat tone. It began by discussing the accomplishments of the partners in the PAFID sector, including the work of the PAFID, the people's organizations, and even the DENR. The letter provided details, for example, of progress made in obtaining stewardship agreements and in ending land conflicts. Accomplishments were portrayed as demonstrative of constructive teamwork. The passage was diplomatically silent, though, about the foot-dragging that occurred in each instance by local DENR staff.

Criticism was in the form of a series of "reservations and suggestions." Three things are noteworthy here. First, criticism focused on the point of difference, never straying into a diatribe against the DENR. The letter set out the perceived problem and explained why the PAFID felt it was a problem. For example, in opposing the expenditure of scarce funds on classrooms in one district, it was argued that the rooms "cannot be used in the foreseeable future because of insufficient population." Second, the section suggested how the LIUCP might be strengthened—notably by providing extra resources for community organizing. Finally, criticism was placed in an ethical context, thereby giving the impression that the PAFID had no choice in pulling out. Criticism was prefaced with expressions such as "we cannot conscientiously endorse," while the letter conveyed concerns that the project was having "an adverse effect on the value systems of the Hanunóo [one Mangyan group]." Such change meant that "the goals of the project are compromised" and that "egalitarian social structures are being adversely affected."

The letter also emphasized that the NGO would seek to ensure that the pullout did not jeopardize either the project or the prospects of the intended beneficiaries. It thus proposed that a memorandum of understanding be signed by the PAFID and the DENR. Such a memorandum would reaffirm the NGO's commitment to the Mangyan, and as part of that commitment PAFID would "continue to assist the DENR and the communities" to promote ancestral rights. The NGO proposed "to provide guidance in income-generating projects without disturbing [DENR] personnel" in a wider effort that would not change "the ultimate goals of the project."

Finally, the letter reiterated the PAFID's commitment to working with the DENR. Although relations over the LIUCP "produced frustrations," the NGO viewed them as "fruitful learning experiences," and it hoped to continue "productive working relationships with the DENR" by building on "past cooperative efforts and learning experiences." The letter reminded Alcala that "much has been accomplished" and indicated that the PAFID believed that accomplishments would be "enhanced" through future cooperation. The importance of the conciliatory tone here cannot be overstated, as the PAFID mobilized to contain the fallout. Employees and PAFID board members used various arguments to defend their decision before DENR officials. Thus the PAFID had "no choice" in the matter, since project deficiencies were souring its relations with communities—something that could not be tolerated, given the NGO's vision and mission. Yet the wish to continue work with the

DENR was reiterated, putting the LIUCP disagreement in a wider context. Indeed, occasional disagreement was a sign of a "healthy" partnership reflective of democratic practices.

The damage limitation exercise appeared to work. As the DENR's Joey Austria (1997) remarked, for instance, there was "no lasting damage," as the saga became "part of a learning process." Indeed, that new work was commissioned only a few months later confirmed that the NGO's stock of moral capital with the DENR had not fallen too severely. Thus, in September 1994, the DENR deputized the PAFID to conduct surveys in Mindoro even as cooperation elsewhere continued (PAFID 1994b). Several factors helped the PAFID. There was the need for the DENR to work with effective and credible NGOs in the post-Marcos era. Then there was the PAFID's long record as a proponent for indigenous peoples. Finally, lobbying persuaded officials that the pullout was indeed consistent with the PAFID's vision and mission—and hence ought to be respected.

Still, there was a sizable financial penalty. The decision meant that the NGO had cut itself off from its leading source of project funding. True, the organization enjoyed some stability due to support from Misereror. And yet there was a substantial shortfall. Staff numbers, which had risen on the back of the LIUCP to fifty, were slashed in 1994, causing much distress. While efforts were made to assist those made redundant, staff levels had not regained their LIUCP peak two years later, when thirty-nine people worked at the PAFID. There was, therefore, a substantial financial and personal cost involved in taking the "moral high ground." On the other hand, the pullout enabled the organization to rebuild relations with Mangyan partners. The formation of the KPLN was a vehicle to pursue land rights and provided a new institutional forum. The PAFID also promoted a Mindoro bill in Congress that provided legal weight to ancestral domain. Demonstrations took place; in October 1996, for example, employees protested at the DENR Mindoro headquarters when President Ramos was there to prod local officials into action. Finally, PAFID staff apologized to Mangyan partners for their role in the fiasco even as they pointed out that they too had been "used by the DENR" (Vargas 1996). In the process, they showed contrition even while emphasizing that both were victims. Constructive engagement was restored and with it the ability to accumulate moral capital in the community—suggested by the PAFID's intensified links to the area in subsequent years.

The LIUCP incident reveals how the quest for moral capital may lead to "irrational" decisions. True, this sort of incident is rare. Less rare, though, are situations in which an NGO can acquire business funding—along with the ethical questions implied by this decision. In general, the opportunity to obtain business funding increased in the 1990s as the business sector prospered under neo-liberalism (Serrano 1994). This opportunity certainly applied to the reform-minded NGOs of concern in this book. Yet partnership with business can raise troubling questions for NGOs. With which corporations should an NGO work? How does cooperation affect NGO criticism of business on social and environmental matters? The case of Haribon helps us to assess the ambiguous position that NGOs can find themselves in regarding business links.[7]

Haribon has longstanding links to business. Indeed, it was initially a forum for wealthy individuals interested in the outdoors (Noering 1982; Roque 1997). Yet, by the late 1980s it was no longer possible to separate environmental issues from development questions, and practice at Haribon reflected this shift (Alegre 1996a). The NGO thus won national and international renown for denouncing corporations and politicians linked to logging as well as destructive practices in the fishing industry (e.g., *muroami*), the energy sector (e.g., Mount Apo geothermal development), the wildlife trade (e.g., turtles, monkeys, seahorses), and heavy industry (e.g., Bolinao cement plant) (Haribon Foundation 1989c, 1990d, 1991b, 1992;

Emmons 1997). Haribon also helped local communities fight businesses and politicians through paralegal training, community organizing, and political lobbying (Barrera 1990; Haribon Foundation 1991c). This effort promoted what Kalaw (1989: 1) called the "democratization of the access to and management of natural resources." Such action also aimed to draw upon indigenous management ideas such as *hanap-buhay,* which focus on "life-flow" (Kalaw 1989; see also Kalaw 1997). This sort of thinking was seen as a precondition for sustainable development even as it necessarily attacked entrenched business interests.

Yet the "custodian of the country's vanishing flora and fauna" (Barrera 1990: 9) also maintained links to a business world sometimes implicated in the depletion of biodiversity. It did so by capitalizing on contacts (social capital) and keen wildlife interest among some business leaders and their families. There is a direct connection to business. Kalaw and Tongson had business careers before joining Haribon, for instance (Lara 1974; Tongson 1996). The board of trustees has also included the likes of Lori Tan (owner of a national bookshop) and Belen King (whose family owns a department store).[8] Corporations have been members and/or made donations, including large firms like the Philippine Long Distance Telephone Company, the Benguet Corporation (active in mining), and the Summa International Bank. The NGO has also been keen to expand corporate membership. As noted, Haribon rejuvenation in the mid-1990s was premised partly on such hopes. In 1996, for example, there was a separate corporate membership (P35, 000 or U.S. $1,346), which included a complimentary copy of a John Ruthven eagle print (Haribon Foundation 1996b). Finally, the NGO has joined with corporate sponsors in fund-raising events. In April 1989, for instance, it hosted with Ayala Land (linked to a leading business clan), the Phinma Group, the Diamond Motor Corporation, and private radio stations a benefit screening of the film *Gorillas in the Mist* (Haribon Foundation 1989d).

Haribon has thus been comfortable working with business. Indeed, one reason for the appointment of Tongson as executive director was to "instill" a corporate mentality in the organization (Tongson 1996). Yet sporadic criticism of business persists. There was the campaign against the Bolinao cement plant, for example. There was also the fight to prevent the end of the export ban on lumber during the 1998 financial crisis, with Tanggol-Kalikasan (1998: n.p.) querying the DENR's ability "to monitor compliance by logging companies to prevent the recurrence of abuses (e.g., illegal logging, inaccurate reporting, tax evasion, etc.)" (see also Beja 1999).

Business connections have raised tricky moral issues requiring the leadership to assess each link separately. One way in which Haribon addressed these dilemmas was by compiling "clean" and "dirty" lists of companies. This mechanism is an informal process by which senior staff members assess whether relationships are appropriate. The purpose here is to establish rules of association that safeguard the NGO's name. According to Tongson (1997), since it was vital that acceptance of corporate support should "not effect the credibility" of Haribon, the organization avoided firms with "bad track records." To do otherwise, he opined, "would kill us."[9] Indeed, there was always the fear that one false step might lead the NGO into being viewed in disrepute, as with the "mutant" NGOs viewed with such distain and concern in the Philippine NGO sector (Aldaba 1992; Constantino-David 1992; Bryant 2002b).

What sort of company makes the dirty list? First, any firm against which the NGO is campaigning is on the list. Public protest, political lobbying, or legal action against a company would warrant inclusion here. It is about avoiding—and being seen to avoid—a conflict of interest. As such, the membership officer consults with

Tanggol-Kalikasan over prospective corporate members so that background checks can be run (Holopainen 1997). The second sort of company on the dirty list is that associated with large-scale resource extraction or heavy industry. Firms linked to disasters are given a wide berth. The exemplar is Marcopper Mining—a company part owned by Canada's Placer Dome, which was blamed for devastation in Marinduque when mine wastes escaped into a river (Haribon Foundation 1996a). The objective here is to avoid becoming beholden in any way to firms involved in environmental "plunder." Firms in this category crop up in the first category as well, insofar as Haribon targets violators (Tongson 1997).

Haribon has certainly used the dirty list. In the 1980s, for instance, it accepted money from a logging firm for a health project in Palawan prior to launching the anti-logging campaign. When that campaign began, though, unspent funds were returned. In the early 1990s, the decision to support Bataan fishers against a firm allied to the Petron Corporation meant that the leadership felt obliged to reject a P40, 000 (U.S. $1,538) donation from Petron (Nozawa 1996). Such action has also involved the rejection of specific individuals linked to "bad" practices. In 1993, for example, Palawan logger Jose Pepito Alvarez met Kalaw and asked to join Haribon, but this request was denied on account of this individual's poor environmental record (Nozawa 1996; see also Broad with Cavanagh 1993).

Haribon also has an informal "clean list"; provided a firm does not belong on the dirty list, it is a prospective partner. The clean list included corporate foundations such as the Ayala Foundation—here reflecting a broader assessment that "gaining access to finance from an 'arm's length' corporate intermediary" was less likely to damage the NGO's reputation (Fowler 2000: 110; see also Anheier and Toepler 1999; Richter 2001). Others featured were service sector firms such as banks, advertising agencies, or fast-food restaurants (Holopainen 1997). Indeed, Tongson (1997) suggested that firms like Coca-

Cola or MacDonald's were acceptable if funds went into a blind trust so that they would not thereafter be able to "use the Haribon Foundation as an endorsement of itself." The clean list is thus for firms or philanthropic foundations that appear to lack a connection to practices that severely degrade the environment, such as natural resource extraction or heavy industry. They are largely urban based and of a "light" service character familiar to the mainly urban-based staff of the NGO. This division of firms into "good" and "bad" categories is certainly part of a wider effort to keep close tabs on the corporate world. There is an eye to consciously countering the "strategic PR" that corporations use to defuse criticism, divide opponents, avoid regulation, and enlist "co-opted" NGOs and their favorable reputations to the corporate cause (Utting 2000b; Rampton and Stauber 2001; Richter 2001).

And yet, there are great difficulties and ambiguities surrounding this entire effort to categorize firms. Perceived inconsistency may arise. To take the Haribon examples noted, the conflict-of-interest rule was used vis-à-vis the Palawan logging firm and Petron Corporation. Still, this begs the question as to why they were not avoided from the start, given their resource extraction credentials. The companies in question, after all, had hardly changed their practices.[10] Doubt can also be cast over the "cleanliness" of firms on the clean list. Indeed, the issue here was summarized in an editorial in the Haribon Foundation (1996a: 2) magazine: "The proliferation of frameworks of thought on environmental activism has begun to confuse the environmentalist. Greenwash is prevalent, thrown in by corporations to make their products appear environmental, even if oftentimes, they are only 2% less threatening to the Earth than before. False or misled prophets abound." While Tanggol-Kalikasan checks prospective partners for conflict of interest, this is an inadequate guarantee of cleanliness. For example, to what extent is it reasonable to include the McDonald's Corporation on either social or environmental grounds (Schlosser 2001)? A similar point applies to the Ayala Foundation—a Philippine business foundation widely seen to be a front for "improving the image" of the Ayala family (anonymous interview). Indeed, Haribon has had links to this dynasty before. In 1984, for example, the NGO held a cocktail party at Manila's Intercontinental Hotel for over one hundred and fifty business and diplomatic leaders to raise funds to support its work. The honorary chairman was none other than Ayala kingpin Jaime Zobel de Ayala (Haribon Society 1984).

The Haribon Foundation does not advertise these lists. There is a clear risk that, were it to do so, it might generate a financially damaging "antibusiness" image, leading to a loss of sponsorship, even as publication of links to business might put off some nonbusiness partners hostile to corporations. In fact the question of business sponsorship provoked fierce debate within the organization itself. Some warned that it might be compromised through alliances to firms wishing to green-wash their images while others thought issues not so "black or white" (Tongson 1997). One Haribon veteran sought "areas of compromise" whereby it would work with natural resource firms that were now environmentally responsible since they had "a moral obligation" to pay for past damage, with Haribon facilitating this process (Plantilla 1997). Such divisions render list making difficult. Yet use of the words *clean* and *dirty* with morally charged connotation suggests a basic strategic tension. There is perennial difficulty reconciling a multifaceted quest for moral capital with recognition that private philanthropy (perhaps "ill-gotten") may be the key to survival. The internal dissent just noted is part of a broader concern about damage to the Haribon reputation if it is seen as "in the pocket" of business (Tongson 1997).

When NGOs pursue moral capital therefore situations may arise in which employees feel the need to reject funding to safeguard a name. Whether withdrawing from projects or refusing business money, NGOs demonstrate that strategizing is about more than cash. The art of fund-raising is partly about knowing when to say no so as to avoid pernicious dependency. One way to address this headache may be by acquiring money from funding bodies managed by Philippine NGOs themselves.

Honor Thy Neighbor?

NGO-managed funding mechanisms involve NGOs in "the setting of the priorities for these funds, as well as their actual management," thereby allowing "a transfer of decision-making power . . . from Northern donors or the national government to Philippine NGOs" (CODE-NGO 1997: 16). That small NGOs participate may have "leveled the playing field" while possibly promoting solidarity through transcendence of "previous ideological differences or past tensions" (Lopa and San Juan 1996: 132–33). There is, then, prospective *collective* benefit through greater autonomy or equitable allocation. But what impact does this have on the quest for funds and moral capital by *individual* organizations? As noted, NGO-managed mechanisms are highly desirable and increasingly important as a funding source for many NGOs. Yet new opportunities place contradictory pressures on NGOs who pursue visions and missions in competitive circumstances.

That organizations compete for funding is a fact of life that engenders much backbiting. This is reflected in the morally tinged criticism of "unfair advantage." Much criticism here boils down to a view that rivals are "not local and/or accountable enough." There is thus tension between international NGOs and large Philippine NGOs. Haribon and the Philippine Rural Reconstruction Movement (PRRM) criticize work by Conservation International and Plan International. These "interlopers" use global networks and cash reserves to "buy" entrance to the country, thereby undermining

Philippine NGOs, who feel "resentment." But there is also tension between large and small Philippine NGOs. Here, it is the turn of Haribon and PRRM to be branded as "Manila imperialists" bent on imposing national priorities on local concerns. They are "too big" to understand local problems and "too preoccupied" with Manila-based agendas. Yet, because large NGOs have track records with the "big shots" (donors and state agencies), local NGOs are passed over despite being "closer to the ground."[11]

Still, funding does not inevitably lead to division. Indeed, the idea for the first NGO-managed environmental funding mechanism grew out of cooperation between the World Wildlife Fund (WWF) and the Haribon Foundation in the late 1980s. Similarly, Haribon maintained links with the Environmental Legal Assistance Center (ELAC) in Cebu City concerning paralegal training (Saniel 1997). For Tongson (1997), this was about "sustaining counterpart institutions" rather than creating a permanent local presence. The debt-for-nature swap involving the WWF, Haribon, and the DENR was the first of its kind in the Philippines. The WWF used funds from Pew Charitable Trusts and USAID to purchase $2 million worth of Philippine debt, dedicated thereafter to environmental conservation. Twenty-three projects were supported, with Haribon playing a leading role. The funding was a boon for the NGO. It used some money for the Mount Isarog project, some for coastal resource management in Batangas where reefs were being destroyed, and some in the battle against aquarium fishing in which fishers used cyanide to catch fish (Pajaro 1992; Haribon Foundation 1997b; Lee 2004).

This funding enabled Haribon to consolidate its leading position in the Philippines. According to Kalaw (cited in Haribon Foundation 1989a: 12–13), the grant was given "because of Haribon's acceptability, credibility and long association with WWF in various environmental projects." Here we see how at a time of transition Haribon benefited financially from moral capital accumulated with a long-standing transnational NGO partner through scientific cooperation. Yet if the grant reflected Haribon's moral capital, it helped to *further* promote the NGO's name. It enhanced Haribon's reputation internationally—here, for a local application (with the WWF) of the debt-for-nature swap first essayed in Bolivia in 1987 (Jakobeit 1996). This provided a solid base both for international fund-raising and political support, as Haribon became a respected actor in the transnational NGO sector—perhaps even sharing in some of the moral capital enjoyed then by the WWF. The swap also led to the elaboration of links to the DENR, the media, and other national actors involved in the environmental sector, as Haribon capitalized on a favorable image as a scientific *and* activist organization (Nozawa 1996).

And yet the debt-for-nature swap program was not an unmixed blessing for Haribon. By the early 1990s it was a source of tension, notably with the DENR, the WWF, and other NGOs. Indeed, as things grew increasingly rancorous, Haribon abandoned a role in a program that it had once led. It is instructive therefore to consider this case a bit further to appreciate the way in which the ebb and flow of moral capital and funding can be linked. The falling out certainly reflected resentment among NGOs (but even some at the DENR) that Haribon had a substantial hold over program funding. Yet other factors were also to blame. Notable here was the way in which Haribon, under the leadership of Junie Kalaw, "hogged the limelight" on environmental issues generally and the anti-logging campaign in particular. Thus the NGO's Palawan campaign received "disproportionate" coverage compared with other pressing campaigns, and it was "domineering" in coalitions such as the Green Forum and the Task Force Total Commercial Log Ban (Green Forum 1993; Legazpi 1994). One NGO leader criticized "a tendency among some environmental NGOs to project themselves as *the* environmental NGO or coalition/federation," raising the fear that there could develop "a monopoly of funds and membership by [a] few major NGOs to the detriment of newly developing ones" (Ganapin 1989: 91). Interestingly, *too much* success at building reputation was arguably linked to a backlash because it was seen to be a threat to the welfare of others in the sector—and thus to the future of that sector itself. Here, then, there appears to be a sharp tradeoff between *individual* moral capital and *collective* moral capital.

Haribon's favorable image also took something of a knock at the international level, at least among selected international NGOs. As Haribon's stature grew, it was approached by a succession of leading international NGOs keen to merge with one of Asia's best-known environmental organizations. Indeed, this process was all about these external actors seeking to acquire greater moral capital in the Philippines by hooking up with its leading national environmental NGO. Both Greenpeace and the WWF approached Haribon about a possible merger at one time or another in the heady early days of the post-Marcos era. Similarly, the U.S.-based World Resources Institute (WRI) was also sizing up the NGO as it sought a Philippine partner.

Yet, each of these possible liaisons ended in failure, generating bad blood in the process. Kalaw's outspoken interventions in the anti-logging campaign quickly put off the politically conservative WRI, while Haribon turned down the other two suitors because it preferred to maintain a separate institutional identity (Nozawa 1996). On the one hand, there was a sense that Haribon recognized its own distinctive "brand value"—something that was likely to be lost in any transnational merger. Indeed, these processes occurred at a time when nationalist feeling was running at a fever pitch over the issue of the U.S. military bases in the country. A merger would possibly make Haribon look foolish precisely at a time when it appeared to have shaken off its image as an "expatriate" organization (*Pilipino Express* 1973; Haribon Society 1982, 1984).

On the other hand, this decision generated predictable disappointment and even resentment among the erstwhile suitors. Indeed, there were local ramifications as the international NGOs made other arrangements. WRI teamed up with the Philippine Friends of the Earth even as the WWF ended up creating its own Philippine branch in the 1990s under the formidable leadership of former Haribon president Celso Roque. In this way, resentment over the failed merger was channeled into competition for funding and moral capital in the 1990s, thus putting new pressure on Haribon's privileged status in the environmental sector.

Meanwhile, Haribon found its reputation under siege from a different quarter when conflict broke out between Kalaw and DENR Secretary Factoran. Relations were already somewhat tense between the NGO and some DENR leaders at the time due to Haribon's role in both the Palawan and national anti-logging campaigns, especially the call for a complete ban on commercial logging (Green Forum 1993). Yet it was Kalaw's vocal denunciation of DENR policies before the U.S. Congress as well as his apparent support for one senator's call for Factoran to resign that sparked the DENR secretary's ire (Haribon Foundation 1990c; Green Forum 1993).

As the feud became highly personalized, there were financial consequences for the NGO. Factoran was joined by Celso Roque and the WWF in pressuring Haribon to "be quiet" about American funding, even as Kalaw was encouraged "to apologize" over the resignation affair (Roque 1997). Roque was particularly incensed over the matter (Factoran 1997). Not only had he brought Kalaw into Haribon in the mid-1980s, but he had also persuaded Factoran that Haribon should lead the debt-for-nature swap program. As a result, Roque resigned from the Haribon board and pushed for the NGO to be dropped from that program. Although Kalaw appealed to WWF leaders in the United States, the NGO had already soured relations with that organization and was thus eased out over 1992–1993 (Nozawa 1997). Here we see how a fall in reputation—notably linked to the actions of Kalaw—led to a cut in funding as political forces within the DENR and erstwhile allies (such as the WWF) shifted against the NGO. Indeed, there can be a ripple effect for, as chapter 4 described, budget cuts to the Mount Isarog project led to a fall in Haribon's stocks of moral capital in the affected local communities. In short, what had once seemed to be a "virtuous circle" of ever increasing moral capital with a widening array of local, national, and international partners had become a "vicious circle" of declining moral capital as personal and organizational resentment built up among partners about what the NGO said and did.

The creation of the Foundation for the Philippine Environment (FPE) as a Philippine NGO-managed grant-making institution on January 15, 1992, was partly a response to the perceived deficiencies of the debt-for-nature swap. A long-term funding mechanism was needed to protect the Philippine environment that would tap the expertise of a wider array of NGOs than under the swap. In a move virtually unprecedented at the time, Philippine NGO leaders assumed control of the new agency and an endowment of about $22 million provided by USAID and the Bank of Tokyo. The FPE subsequently used the interest income to support NGO projects and administrative costs (Tan 1997).

Experienced NGO leaders with close links to the DENR led the agency. Factoran was the first chair of the FPE board (having stepped down as DENR secretary) and Ganapin became its first executive director (also with DENR experience). A governing board held ultimate responsibility for funding decisions but received advice from regional advisory committees and an experts panel. The FPE was thus "to strengthen capabilities and roles of NGOs, POs and communities" while providing "sustained financial support to organizations responsible for implementing biodiversity conservation and sustainable development programs" (FPE 1995: 11). By the mid-1990s, the annual disbursement was between P24.7 million and P28.6 million—about $1 million per year (FPE 1996: 34).

The FPE certainly widened the pool of recipients, with resources going to an array of NGOs and people's organizations. Both the PAFID and Haribon obtained funding. Both received money in 1996, for example, to conduct training in the Biodiversity Conservation Capability Building Support Project. Other grants have been won. Yet the distribution of funding has emphasized if not reinforced the relative decline of Haribon as a "giant" among environmental NGOs. In 1994, for instance, it received one "responsive grant" worth P975,800 ($37,531) out of a total disbursement in this category of P24,236,003 ($932,154) (FPE 1995: 30).

What are the implications here for NGOs keen for funding and moral capital? The FPE has alleviated but not eliminated the complaint that donors favor established NGOs. True, the push to widen the net provided funding to small and medium-sized NGOs and people's organizations. Yet the criteria of eligibility required applicants to be legally recognized and "continuously operating for at least two immediately preceding years" with a proven track record—"other programs or projects of a similar nature that it has managed with success" (FPE 1996: 39). Thus the existence of the FPE diminished the importance of Haribon, but the need for an established reputation at the same time sometimes barred new entrants.

There is also more of a need to accumulate moral capital in the NGO sector itself. Even though NGOs have long needed to pay at least some attention to what peers think about them, the NGO-managed funding mechanism added an edge to this process since NGOs now needed to compete with *and* judge one another over funding matters. Given scarce funding—in 1996 FPE received three hundred proposals but approved only fifty of them—it is hardly surprising that these funding bodies are plagued by infighting as organizations jockey for funds and influence.[12]

The Haribon experience over the Mount Isarog project is revealing here. As noted, this project was funded through the debt-for-nature swap and supported Haribon's work with four people's organizations that it helped to create. However, as part of the political fallout over the debt-for-nature swap, the FPE assumed overall control of the Isarog project. It thereupon insisted that funding be reallocated among members of a new Mount Isarog NGO-PO consortium. Haribon objected to this move on two grounds. First, while it was in the consortium, its budget was cut in half to fund the others. The only alternative, though, was to leave the project. Second, Haribon was critical of some members of the consortium, especially Plan International, complaining that it was associated with the eviction of settlers from nearby Bicol National Park. Yet with Factoran and Ganapin at the helm, the FPE—in the words of one Haribon employee—was "doing NGO politics" as it rewarded NGOs such as Plan International at the expense of Haribon. Still, the Haribon leadership adopted a "low key approach [that] did not confront FPE over the changes"—thereby acknowledging their weak position and wish not "to lose face." The funding was thus redistributed on a fifty-fifty basis—with the resulting Haribon budget cuts noted above.[13]

It is only to be expected that NGOs become embroiled in conflict over FPE funding. There is, after all, a wider history of conflict over funding linked notably to issues of turf and personality. The fact that NGOs themselves now allocate funding does not mean that competitive pressures have eased. Nor does the rise of these mechanisms as well as their desirability as a source of funding prevent *some* NGOs—especially those with members on the board or on the regional committees—from having disproportionate say in the definition of reputable conduct. In this manner, the search for cash and moral capital through the medium of NGO-managed funding arrangements such as the FPE are likely to continue to be the basis for interorganization tension as charges of favoritism color funding deliberations.[14]

Just as strategizing is rarely reducible to a dash for cash, how NGOs raise money has a direct bearing on the building of a good name. We have seen that the pursuit of moral capital is complicated by financial exigencies that introduce ambiguity: the fear of receiving "some form of 'Judas money' " runs deep as organizations are drawn into tradeoffs not to their liking (Ganapin 1989: 96). Still, "insightful agility" is not synonymous with acceding to whatever is demanded by donors. The preoccupation with financial sustainability is thus not simply a knee-jerk reaction to shifting economic circumstances. It is also an attempted declaration of independence from donor dependency. Hence the seemingly irrational effort devoted to "marginal" fund-raising: membership drives, special events, or promotional products. This attempted declaration of independence is also to be seen in the rejection of funding and the compilation of dirty lists—an exhilarating if risky defiance of powerful actors. At the level of the NGO sector as a whole, there is, too, the promotion of NGO-managed funding mechanisms—although this may not be as liberating for individual organizations as might at first seem. However, the thread running through these activities is a multifaceted quest for moral capital whereby the ability to act is partly conditioned by the reputations that NGOs acquire or lose for pursuing moral purposes altruistically.

The raising of funds is a process whereby moral capital is "translated" into economic capital—it is one "pay-off" for a good track record. Yet moral capital is accumulated or dissipated through the act of fund-raising itself. Just as the quest for moral capital has financial implications, so too the pursuit of money has ramifications for the reputation of a NGO. Still, if the financial strategy that an organization pursues says much about what makes it "tick," this can also be true about the ways in which an organization's vision and mission is mapped across space. To the extent that NGO activities assume a spatial form through campaign or project work, questions are thereby raised as to how they may pursue moral capital through territorial strategies. This is the subject of the next chapter.

CHAPTER 6

Mapping the Mission

In addition to political and economic action taken by NGOs in the quest for moral capital, their function as spatial and territorial actors also relates to that quest. Little considered in the literature, this aspect is vital to a rounded understanding of NGO practice. Where do NGOs undertake their work and why do they make these choices? Are there "no-go" areas, and if so, why are they off limits? How do NGOs interact spatially? Are there patterns in locations that are reflective of territorial strategizing? How do these concerns relate to moral capital?

To the extent that NGOs are moral entrepreneurs, it may also be that they are territorial agents (Johnston et al. 1994: 573–75).The concept of territoriality has come far since the days when it was seen as innate human conduct. In recent times, it has come to be seen as a contingent social and political outcome. Thus Sack (1986: 5) defines human territoriality as "the attempts by an individual or group to affect, influence or control people, phenomena, and relationships, by delimiting and asserting control over geographic area. . . . [It] is not an instinct or drive, but a rather complex strategy . . . [and] the device through which people construct and maintain spatial organizations." It is thus "a primary geographical expression of social power" (Sack 1986: 19). It is embedded in social relations as well as constituting them. There are "tendencies of territoriality." There is a classification of space by area as "ownership," and there is the use of borders to delimit territorial control and power over a prescribed space. Finally, there is territoriality as a means by which power relations are reified.

Much work in political geography even today understands territoriality as an attribute of states. The literature explores the impact of interstate relations on territory and how it is understood (e.g., cold war, European Union), the social construction of identity through "imagined communities," and the possible dissolution of identity and territory through globalization (Walker 1990; Winichakul 1994; Vandergeest and Peluso 1995). This literature also tends to think about territory as a geographically unified or coterminous phenomenon. The state as a "container" in which national space is delimited is the exemplar. This view combining a state-centric perspective with a geographically coterminous view is one I shall call "hard territoriality" (Taylor 1994; Johnston 2001).

Recent work challenges these assumptions, leading to a contrasting view of what I term "soft territoriality." The emphasis is on understanding territoriality as a strategic and relational practice by a range of state *and* nonstate actors operating in a world of complex multiscale activity (Swyngedouw 1997; Kelly 2002; Kurtz 2003; Routledge 2003). For example, such work considers the micro-level territorial behavior of groups and individuals linked to race, class, and gender. A unifying theme is that territory and boundaries are constructed and contested "in peoples' everyday lives" and are a "spatial reflection of power" (Storey 2001: 6). While these processes are "less obvious" and possibly "more difficult to detect and observe" than state-centered processes, this "does not make them any less real" (Storey 2001: 146).

Even this literature, though, tends to view territory as a geographically coterminous phenomenon—albeit at the local community level (Martin 2003). Yet territoriality can be unpacked further to account for behavior by nonstate actors in a geographically diffuse manner (Yeung 1998; Cox 2003). This enables us to consider "nonlocal" and nonstate actors—such as national or transnational businesses or NGOs—in relation to territorial behavior.

Defining Territorial Reach

Diverse considerations come into play when NGOs choose where to work and build their names. These considerations may reflect how the mission is articulated, since there are territorial implications associated with them. Thus there is usually an area-based slant to the concerns of most organizations: upland development, forestry, coral reefs, or urban poverty, for example. The possible effects of such "predisposition" are clearer when the PAFID and Haribon are considered.[1]

The PAFID's territorial ambition is allied to promoting indigenous peoples' rights. Thus, the mission declares it will "make innovative technical assistance available to tribal communities" and "promote holistic community development efforts which are culturally sensitive and ecologically sound" (PAFID n.d.: 2). It will also "advocate in public and private institutions, policies and programs beneficial and respectful of tribal Filipino[s]," thereby facilitating a vision of them "as responsible stewards of resources" (PAFID n.d.: 2).

This mission has various spatial and territorial implications. Since it assumes that indigenous people are socially and environmentally responsible in regard to local resource rights, the PAFID inevitably tries to safeguard those rights *in situ*. There is a strong moral component to this spatial understanding. "Tribal Filipinos have always occupied the mountain areas and their environs since time immemorial. Their ancestors lived and died there, hence the term 'ancestral lands.' They moved around unhampered anywhere in their domain: gathering food, hunting, and later on planting to meet their needs. These mountain people believed, until now, that they belong to the land, and the land belongs to them. Historically as well as morally, they are correct" (PAFID 1993a: 2). In helping to define and defend the territory of indigenous partners, the organization is involved in a "conservative" spatial strategy inasmuch as its area of operation is confined to current indigenous geographies. As noted, those geographies reflect that indigenous people are "driven away from their ancestral lands by virtue of documents acquired from government agencies and backed up by military force" (PAFID 1993a: 2). The strategy is thus to help communities repel or modify intrusions such as logging, mining, or tourism. As board member Donna Gasgonia (1997) observes, the NGO has "no choice but to be reactive" since "the nature of the problem" it faces "requires that approach." In the process, territorial ambitions of the organization are indelibly linked to the contingent geographical status of partners.[2]

What does this mean in practice? Clearly, work is based in "remote" areas where indigenous communities survive. That these areas are far away from offices in towns and cities means a grueling lifestyle for employees, who travel long distances over rugged terrain to get to communities. Physical fitness is essential for appointment, and employees must have a current medical certificate on file (PAFID n.d.: 3–4). It is, as the PAFID's Dave De Vera (1996) put it, a "tough niche" to work.[3]

A more precise sense of spatial and territorial reach is gained by considering work done in the field. In pursuit of its mission, the organization does community organizing, land tenure work, surveying and mapping, political advocacy, infrastructure development, agroforestry, and training (PAFID 1995a, 1995b). There is a strong responsive element here, as activities emerge from detailed discussion with partners and organizational threat assessments.

They also result from unsolicited pleas for assistance from those familiar with the NGO through word of mouth. Such flexibility is vital: "PAFID's basic strength is its availability to go to the field when badly needed and its ability to organize and mobilize communities to counter threats and acts of demolition or eviction and other forms of oppression meant to dislocate people from their ancestral homes. Some of the problems and cases brought to PAFID by tribal people were practically hopeless and were outside its parameters of assistance. However, PAFID has to attend to these people in one way or another because, as investigated, these groups were urgently in need of assistance with no one to assist them" (PAFID 1995a: 4, 7). A varied *spatial* profile is the result. Figure 6.1 notes where the NGO has worked, though it does not differentiate between areas of short and long-term involvement. The latter involve work to secure land title and/or sustained advocacy, since both require much staff time—indeed, serving as a useful indicator of priorities and territorial remit. Accounting for these labor-intensive priorities, then, a *territorial* profile is also provided with a tighter focus in evidence: north-central Luzon, Mindoro, Palawan/Panay, and south-central Mindanao.

This profile illustrates how territorial behavior is linked to area-specific reputation building. Home turf is northern Luzon, with a high profile predictable there. The leadership is largely from this area, including the De Vera family, Willy Tolentino, Sammy Balinhawang, and Pastor Delbert Rice. The "spiritual" center of gravity is in Imugan (Nueva Vizcaya), where Rice has long worked with the Ikalahan and their Kalahan Educational Foundation (KEF). Indeed, the KEF was the first indigenous people's organization in the country to win a twenty-five-year forest land lease from the Bureau of Forest Development (Rice 1996; Anselmo Roque 1997). The "Ikalahan model" became the guiding referent for the PAFID. In September 1977, its board met specifically to discuss the wider applicability

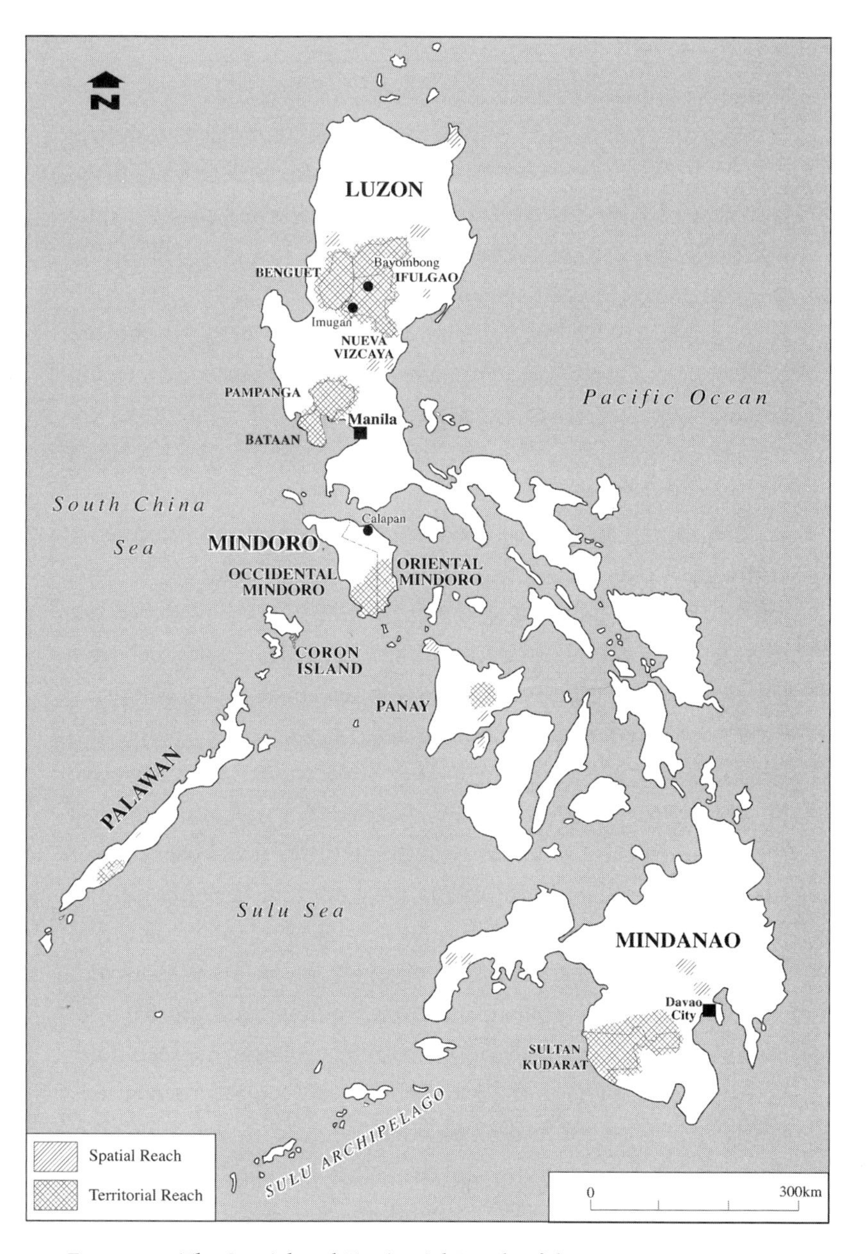

FIGURE 6.1 The Spatial and Territorial Reach of the PAFID, c. 1993
Adapted by Roma Beaumont from PAFID (1993b): 3, 9.

of the KEF lease. From this and other meetings, a role as a go-between in promoting a "problem-oriented" approach to indigenous rights was confirmed. Attention thereafter was given to problems faced by various groups in Luzon as the work of the PAFID became known regionally. Sustained political advocacy was used in the case of the Tinggians of Abra as the NGO joined others in the battle against the Cellophil Corporation (see Chapter 3). In the case of the Aeta of Bataan and Pampanga, the PAFID helped them obtain CFSAs while also providing leadership training and emergency assistance following the Mount Pinatubo eruption (PAFID board minutes 1977–78; PAFID 1995a).

The combination of personal contacts both in and out of government as well as a growing reputation for effective work based on the Ikalahan model meant that the PAFID soon branched out from its home turf. A more complex pattern of territorial behavior emerged. By the mid-1980s it was the leading NGO advocate of CFSAs, and there were new territorial clusters in south-central Mindanao, Panay, Palawan, and Mindoro. There was generally a twofold dynamic here. When the PAFID was a fledgling organization, it fell to Rice and board members to promote it through a network of contacts that included local and national officials, academics, donor representatives, NGO activists, and community leaders. This networking often resulted in an invitation for the PAFID to meet with indigenous leaders—as happened, for example, with the Batak of central Palawan, the Hanunóo Mangyan of Occidental Mindoro, and the T'boli of southern Mindanao. Such contact often turned into a PAFID-led effort to obtain land tenure. This was so in Mindoro, where efforts to transplant the Ikalahan experience resulted in two applications for the Mangyan in 1980 and 1982. It was also true in Mindanao, where a communal land title was filed on behalf of the Manobo (related to the T'boli) in 1987. In the mid-1980s the NGO was similarly active in Palawan, with four CFSAs filed on behalf of

Palawano and Tagbanua communities. Finally, a foothold was established in north Panay when the organization assisted an Ati community in filing for a CFSA (PAFID 1993b: 3).[4]

Yet the PAFID also needed to undertake localized reputation building on its own to solidify a presence in these communities. For example, work in one community led the Tagbanua leader of Coron Island to contact them, as noted in chapter 4. It is in Mindoro that the connection between territorial behavior and local reputation building is most striking. Following the tenure applications noted above, links to other communities were forged in Oriental Mindoro. The key to influence here was a growing reputation as a "friend of the Mangyan" in their struggles. Thus, for example, board member Julian De Vera was approached in 1984 by the barangay captain of Arigoy-Luyos "for help against harassment and intimidation" by the owners of a tree plantation and their allies in the military (PAFID 1984: 5). De Vera offered "unwavering support of the Mangyan in the area," providing advice on legal rights and the creation of a people's organization. This interaction—repeated in most communities where the organization works—provided much-appreciated moral and practical support at a critical moment. This was "repaid" through the buildup of moral capital.[5]

Requests for assistance thereafter flooded in. Much work with communities was about preparing them "for battle" with cattle ranchers and other opponents (De Vera 1996). This was a complex, even dangerous process involving official petitions, community organizing, public protests, and negotiations with local bosses. These activities were at the heart of what staff members did as they promoted the Mangyan cause. I return to this point below. Here, it is important to appreciate how this local track record was intimately associated with an expanding territorial base there. Successful community project work led to requests for further help from a community—help with livelihoods once a CFSA was obtained, for example.

It also led to requests for help from other communities. Thus land tenure had been negotiated on behalf of eight Mangyan communities by 1992, with many others receiving support of some kind (PAFID 1993b). By the mid-1990s, the PAFID led a national campaign for Mangyan rights with a "significantly increasing number of organizations . . . asking for PAFID's assistance in land tenure, surveys, and organizing" (PAFID 1995b).[6]

The PAFID is not a "national" NGO in the sense of having sustained operations in all regions. This is partly due to a strategy of creating a few territorial clusters in which the NGO is preeminent, though this approach does not preclude work elsewhere. Nor is it static, as there is a need to be flexible in new political or economic conditions. There are two advantages to this approach in terms of moral capital. It has been a felicitous means to extend the PAFID's reputation for "doing good" locally. This area-based reputation is also important for donors, state agencies, or the media. For example, donors speak admiringly of "grounded" NGOs "able to integrate with locals" and "connected to communities" (Braza 1997; Chua 1997; Racelis 1997). If a track record is partly about accumulating experience in specific localities, then a territorial strategy can help acquire such experience by enabling "spatial economies of scale" in work.

Territorial ambitions are closely associated with biodiversity conservation for the Haribon Foundation. This organization shifted from an "environment-first" approach to one that integrated social and environmental concerns such that, since the late 1980s, it emphasized "conservation through people" with "a particular emphasis in the empowerment and participation of local communities in conservation" (Haribon Foundation 1997a: 1). Indeed, for biodiversity conservation to succeed there needed to be a "critical mass supportive of conservation" (Haribon Foundation 1997a: 1). A twofold strategy was required: the NGO identified endangered biodiversity "hotspots" and then sought a local constituency through projects integrating conservation with livelihood. Such action derives from a Haribon vision to ensure ecosystems are "conserved and managed in a socially and scientifically sound manner" by involving "communities that are environmentally aware and responsible stewards of the environment"—all with an eye to "equitable [resource] access" and "quality of life" for residents (Haribon Foundation 1997a: 7). A key aspect of the associated Haribon mission is to pursue "community-based resource management strategies in specific sites" (Haribon Foundation 1997a: 7).

The Haribon vision is reflected in its work, including research, community organizing, paralegal training, livelihood support, litigation, advocacy, and education. These activities take employees to many regions. As with the PAFID, a wide spatial profile contains within it a narrower territorial profile, in that the organization sustains presence in selected areas only (see fig. 6.2). A focus has been central Luzon, with this "home turf" covering an area reachable through a one-day car journey from Manila. Many employees hail from this area and typically went to school and university there as well. Thus there is a concentration of school and university chapters stretching from Central Luzon State University (Nueva Ecija) in the north to the University of the Philippines, Los Baños (Laguna), in the south. True, the activities of students and employees extend beyond the area. Still, Haribon projects are nonetheless prominent in central Luzon. They include a marine conservation project at San Salavador Island (Zambales), a coastal resource management project at Bolinao (Pangasinan, as discussed in chapter 4) as well as at Anilao (Batangas), and net-use training in various communities in Batangas, Pangasinan, Quezon, and Zambales (Lee 2004).

If central Luzon is home turf, the organization is also well known in two other territorial clusters. The first is in Camarines Sur (south Luzon) at Mount Isarog. As noted, activities revolved around the

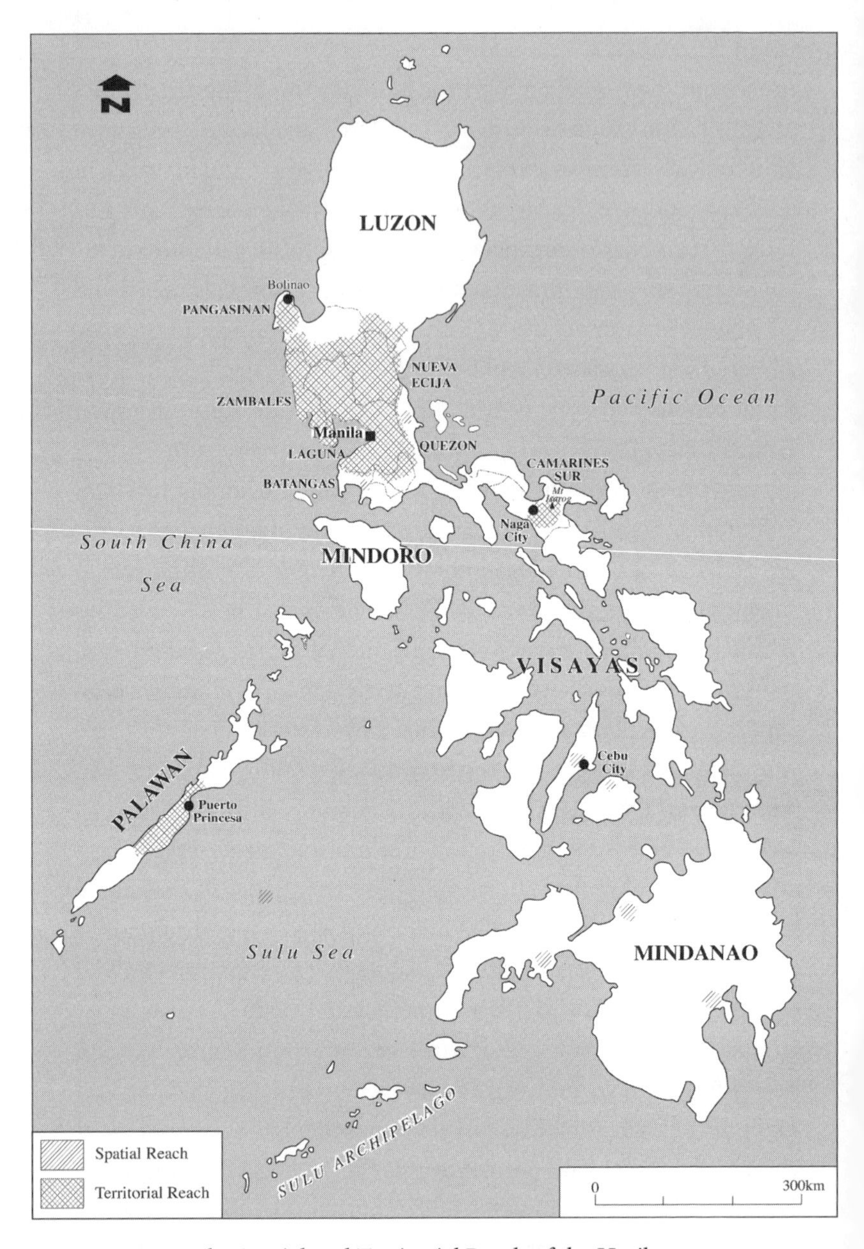

FIGURE 6.2 The Spatial and Territorial Reach of the Haribon Foundation, c. 1996
Adapted by Roma Beaumont from Haribon Foundation (1997a): Annex D.

National Park, with additional support provided by the student chapter in Naga. The latter conducted the baseline survey in 1989, and members thereafter pressed the local government and the DENR on illegal logging. Prominent action by Haribon Ateneo de Naga facilitated community work by Haribon project officers in the 1990s inasmuch as moral capital accumulated by the former helped facilitate the entrance of the latter to the area (Cabague 1991). Overall, the combination of a student chapter and project work meant that the NGO developed strong links and a good "base" of moral capital locally.

Haribon's work in Palawan best illustrates its territorial behavior. That Palawan is a final forest frontier rich in endangered biodiversity explains entry (Clad and Vitug 1988). Following a national signature drive in 1988 under the banner "Boto sa Inangbayan" (Vote for the Motherland), Haribon Palawan was established under the direction of attorney Joselito Alisuag. This became the leading group of environmental activists in the province—a middle-class group of tourist guides, activists, government officials, divers, and lawyers (Broad with Cavanagh 1993; Arquiza 1996). In addition to the fight against logging, Haribon Palawan conducted high-profile battles against mining in the Irawan watershed near Puerto Princesa, seaweed extraction in Tubbataha Reefs National Marine Park, and destructive *muroami* and *pa-aling* fishing (Haribon Foundation 1989a, 1990b, 1990c, 1991a; Barrera 1990; Connell 1991). This work attracted an oppressive response (Lopez 1991; *Philippine Daily Inquirer* 1991a, 1991b). Political advocacy was used to help a green candidate, Edward Hagedorn, win the mayoralty race in Puerto Princesa and subsequently assist him in clamping down on illegal activities. Evidence of territorial thinking is also evident in community-level work such as a European Union–funded nontimber forest project and an ancestral domain initiative with a Palaw'an community in southern Palawan (Haribon board minutes 1996; Lozada 1997). The latter built on prior activities—in 1989, for instance, when Haribon assisted a community in north-central Palawan to form a people's organization and obtain a communal forest lease (Haribon Foundation 1989c). These varied activities highlighted a local reputation for promoting community-based alternatives to resource overexploitation.

It is interesting that Haribon's "name projection" is greater than its record on the ground might suggest. This reflects a strategy of building international research and advocacy links (Tongson 1996). It is linked to training courses on law and ecology that involve visits to communities nationwide (Luna 1996). Yet the Haribon reputation also derives from a territorial strategy involving biodiversity conservation in selected areas. Palawan stands out here, but central Luzon and Camarines Sur are other clusters in which reputation is built. These clusters are not an overriding focus of action, but they are indispensable to an organization criticized for not being "sufficiently grounded" due to political advocacy (Romero 1996). Territorial behavior here is not simply a means to obtain spatial economies of scale in reputation building. It is also a counterbalance to other activities at the national and international levels such as political advocacy or conference work.

The spatial and territorial behavior of the PAFID and Haribon is thus partly about mapping organizational mission. It is also an attempt to boost moral capital. The development of area-based reputations is integral to generating track records critical to NGO empowerment. As one observer put it, "rootedness" was the key if credibility was sought (Racelis 1997). And yet, the experiences of these two NGOs show differences in the way in which territorial strategy is articulated. While the PAFID is primarily focused on a mapping process about indigenous communities, Haribon is first and foremost involved in a mapping strategy dedicated to ring-fencing biodiversity with concern about local people ancillary to that primary concern. This difference certainly can lead to different expectations and tensions surrounding the quest for moral capital. In comparison to "development" NGOs like the PAFID, environmental NGOs like Haribon often have a tougher time in convincing local residents that livelihoods matter as much as (let alone more than) the protection of endangered flora and fauna. That said, these differences suggest more the need for finely tuned strategizing by organizations reflective of vision and mission than a sense of obstacles to capital accumulation per se.

Whatever the rationale for territorial strategizing, there are certainly institutional implications of such behavior. For "national" NGOs, there is a struggle to reconcile local responsiveness with organizational cohesion, and this is reflected in tensions over centralization and decentralization. These tensions were enhanced as a result of wider political change. Thus a key trend in the 1990s was the devolution of selected powers from Manila to local government under the 1992 Local Government Code (Osteria 1997; Clarke 1998). For the PAFID and Haribon the need to be—and be *seen* to be—locally responsive has provoked interest in decentralization.[7]

Haribon has long operated in a quasi-decentralized manner. Although projects tend to be coordinated from Manila, local membership chapters provide room for autonomous action, including campaigns and projects. Haribon Palawan is noteworthy here. It was created as a regional chapter with its own funds and agenda—the aforementioned EU nontimber forest product project being a case in point. Such autonomy was not viewed as an unmixed blessing in Manila. On the one hand, this arrangement satisfied the need concerning local responsiveness, thereby providing a basis for moral capital accumulation with various partners around the country (not only in Palawan). On the other hand, though, it raised the problem of who controlled the Haribon name and hence what was done under its aegis. This was viewed as a problem by staff in Palawan as well as staff and board members in Manila (Haribon board minutes

1990–1996; Nozawa 1996). Indeed, one of Tongson's first tasks upon being appointed executive director of the national organization in 1995 was that of "fixing the management structure" relating Manila to the chapters (Tongson 1996). Yet the effort to tighten central control provoked resistance in Palawan, thereby underscoring the difficulties faced by large NGOs in getting the balance right. Here, then, is a case in which different levels of the *same* organization want the moral capital of the NGO for their own purposes, perhaps even in ways that are not satisfactory to the other level.[8]

The PAFID chose to decentralize management in the early 1990s, with three units established to serve Luzon, Mindoro (including Panay and Palawan), and Mindanao. These units had their own staff and offices located in a key provincial town and were run by experienced employees with "leeway to operate semi-autonomously" (PAFID 1995a; De Vera 1996). For example, Sammy Balinhawang ran the Luzon office from the town of Bayombong, where the NGO had long-standing links. This was useful both because of the town's strategic location and because of preexisting moral capital enjoyed locally.

And yet there were pros and cons to this structure. On the plus side, it permitted staff to respond flexibly and swiftly to events on the ground. The *symbolic* importance of the offices is not to be gainsaid either, since they signaled to indigenous communities that the PAFID was "there to stay." Partial decentralization thus strengthened the ability to accumulate moral capital through area-based action. "This move provided a certain geographical focus in the organization's immediate response to a community's concerns and in the accomplishment of targets. A lot of positive responses have been received by the organizations [*sic*] regarding this move. The number of POs/NGOs asking for PAFID's assistance is growing and may serve as a testament to the PAFID's efficiency, competence, commitment, and credibility" (PAFID 1994c: 4). Here again, though, there was a reduced ability to control use of the PAFID name nationally. This was a problem in at least one instance in 1996 when an employee in Davao wrote to a colleague about DENR "corruption." This note was seen by Joey Austria, who resigned from the board in anger, depriving the PAFID of the support of a trusted ally. In the end, he was persuaded to return, with no lasting damage (Rice 1996; Austria 1997). The point is that this "gaff" would probably not have occurred in a context in which there was tight central control of name and moral capital.

The link between moral capital and territorial behavior is also seen in NGO hiring practices. As noted, local residents formed Haribon Palawan, as with other chapters. Here, reputation is acquired for work done by *local* people, thereby undercutting charges of "Manila imperialism." To some extent the same can be said about the PAFID inasmuch as work is usually done by staff from the area. For example, Willy Tolentino (1996) has led advocacy in relation to his Kankanai people in Busol (Benguet) as this group has faced intrusions on its lands. Once again, though, there is a tradeoff between enhanced moral capital through local "rootedness" and the greater risk of damage to the PAFID's standing nationally if that capital is then used in ways at odds with a vision and mission established in Manila (as purportedly happened in northern Luzon in the late 1970s; see chapter 3).

Personal qualities can certainly boost moral capital and strengthen territoriality. At Haribon, community organizers often have a background similar to that of those with whom they work—two of four employees at Bolinao in 1996 had been fishers before joining the NGO, for example. Such experience can help build trust and friendship. At the PAFID, meanwhile, it is standard practice to hire indigenous people—twenty out of forty employees were indigenous in 1996. Indeed, one of them, Willy Tolentino, was executive director. There were clear benefits here in boosting the PAFID's local appeal. For Rice , work was easier because indigenous staff "facilitated empathy." For Dave De Vera, this process "added a sense of legitimacy" to what the PAFID did. Nevertheless, relations between indigenous staff and indigenous communities are not always harmonious. Since many of the former hail from northern Luzon, cultural differences between different indigenous communities have arisen. This was so at Coron Island when a community organizer from Luzon fell out with Tagbanua residents over a livelihood project. Overall, though, the hiring of indigenous employees has stood the organization in good stead by solidifying operations and serving as a "calling card" for new contacts—and hence yet another aspect to the accumulation of moral capital.[9]

There are, in sum, important implications that follow for the PAFID and Haribon from the interlinked pursuit of moral capital and territory. This is a complex process riddled with ambiguity. Such complexity partly reflects a wider context characterized by multiple and overlapping territorialities.

Contending Territoralities

Nongovernmental organizations are usually keen to defend territorial prerogatives since area-based reputations can be crucial to their empowerment. Reputations are won and lost partly on an ability to attain spatial economies of scale through work clusters. How far NGOs go to defend territory—forgoing revenue or risking employee lives—speaks volumes about the importance to them of maintaining a reputation for area-based work. Take, for instance, the experience of the Haribon Foundation. The "name projection" of this organization is partly attributable to the way in which it stood its ground in Palawan even though under intense pressure to leave. As noted, Palawan provides a good example of NGO territorial behavior. Such behavior took place against a backdrop of persistent intimidation that strengthened the resolve to continue work in the province.

When Haribon became active in protecting the forests of Palawan, formidable political and economic barriers hindered action. The political economy of natural resource "plunder" implicated Palawan politicians (including House Speaker Ramon Mitra), making for a "very oppressive" climate even after Marcos (Nozawa 1996). It was a risky proposition for an NGO hitherto known for scientific research and nature walks. Yet the need for action was evident. In October 1987, for instance, Haribon held a bleak meeting on Palawan logging and deforestation (Haribon Foundation 1987). The campaign began the following year as field activists and Manila staff met clandestinely to avoid coercion. The local chapter was formed to defend against the charge that the NGO was "foreign" to the province—thereby underlining the connection between strategy, "rootedness," and moral capital. In addition, the Manila office mounted a national campaign to save the Palawan forests, gaining wide media coverage.

But the NGO thereby earned the enmity of powerful opponents. In Palawan, logging kingpin Jose Pepito Alvarez used his network—including mayors, the governor, and two congressional representatives—to harass Haribon. The harassment included physical intimidation and death threats. While the local media controlled by Alvarez allies vilified the NGO, politicians sought to ban Haribon's Manila staff from visiting the region. Politicians in Congress denounced the organization. Indeed, the latter was forced to appear before the House of Representatives Committee on Natural Resources co-chaired by Palawan's David Ponce de Leon. According to Kalaw, the hearing was "an orchestrated effort to pressure Haribon into stopping the campaign" (cited in Haribon Foundation 1989a: 13). The NGO intensified its efforts through the Task Force Total Commercial Log Ban even as Kalaw publicly rebutted charges made by De Leon and Mitra.

In Palawan meanwhile, a Haribon member precipitated a political crisis when he stumbled across a hidden cache of contraband logs belonging to the military. The draconian response was the arrest of fourteen Haribon members, including Alisuag, under the anti-subversion law. According to the military, communists had "successfully infiltrated" the NGO using environmentalism "to gain mass support" (Haribon Foundation 1991a: 4). Kalaw rejected the charges and accused Mitra of orchestrating the arrests to prevent further revelations about illegal logging. Haribon threatened to file charges against the Philippine National Police and the armed forces for wrongful detention. In the event, the case soon collapsed and a meeting between Haribon, the armed forces, and the Philippine National Police in March 1991 defused tensions. In the wake of the crisis the NGO maintained a high profile in its battle against logging in Palawan.

As Cristi Nozawa (1996) observed, this battle was a "coming of age in political advocacy" for Haribon. The battle was also a ringing declaration of *territorial* commitment insofar as it indicated that the organization would not leave the province, as virtually every powerful political, economic, and military figure wished. It was an opportunity for Haribon to "show its mettle," thereby earning moral capital with donors, journalists, academics, and others opposed to logging. Such courage reflected, above all, a territorial campaign that staked Haribon's claim to lead in Palawan. In the process, Haribon defeated the challenge from opponents in local government, the DENR, and the logging industry that had their own territorial ambitions. Here is a graphic indication indeed of how tenacious an NGO can be in defense of territory, with attendant benefits in terms of moral capital.[10]

The PAFID has shown a similar tenacity. In Mindoro, for instance, the employees braved threats of physical coercion and the loss of funding to maintain operations. Here, too, territorial behavior is linked to moral capital even as reputation building itself reflects area-based commitment. One challenge in Mindoro was financial. Recall the events surrounding the Low-Income Upland Communities Project (LIUCP) discussed in chapter 5. The PAFID and others had pulled out in late 1993 due to tension between NGOs and Mangyan partners. I noted that the PAFID was prepared to sacrifice much to remain on good terms with communities. Here, I describe how deep local roots resulted in an *ongoing* commitment to nurturing links and an area-based reputation there.

The PAFID's wish to stay put was affirmed in the letter of termination to the DENR secretary in December 1993. Yet good intentions are vulnerable when there is financial hardship. Here, though, the PAFID record in Mindoro *after* 1993 highlights that the quest for moral capital through territorial behavior can be an inducement to stay despite the costs. Following the wave of redundancies, a much-reduced staff was maintained in Mindoro with salaries paid from a Misereor grant and the NGO's meager savings. During 1994, employees worked hard to rebuild trust and to advance land tenure for indigenous communities. They supported the Kapulungan Para sa Lupaing Ninuno (KPLN), which had been established in 1993 as a vehicle to promote ancestral claims. Dave De Vera and others, led by the board's Delbert Rice and Donna Gasgonia, pushed the Mindoro bill in Congress. Meanwhile, efforts to obtain new funding paid off in 1994. Misereor confirmed a grant to support advocacy and land tenure work following a glowing independent assessment (PAFID board minutes 1994). A grant from the Biodiversity Support Program (BSP) followed, providing support for a survey and mapping unit aimed at helping to map ancestral domain. Critical here was acquisition of state-of-the-art Global Positioning System (GPS) equipment enabling rapid mapping (PAFID 1994b).

The NGO thus geared up again in Mindoro, with staff numbers reaching seven full-time employees by October 1996. Local partners were trained to use GPS equipment and acquire other skills using materials "which utilized Mangyan concepts and characters that facilitated easy and effective learning by the target users" (PAFID 1994c: 7). The objective here was compilation of evidence for a CADC application, and by February 1995 six applications covering some 170,000 hectares were nearing completion with PAFID help. As word spread, the organization was in ever greater demand, and by the mid-1990s it led the islandwide effort to document Mangyan claims. This experience illustrates two things. First, it provides further evidence as to the willingness of NGOs to sacrifice in order to remain in an area. As with Haribon in Palawan, PAFID clearly did not intend to abandon its base in Oriental Mindoro despite the LIUCP setback. Second, it illustrates that this territorial strategy can result in subsequent payoffs—strengthened local links plus moral capital *and* new funding opportunities based on a solid area-based reputation.[11]

There were other tests of PAFID's determination concerning Mindoro. It was common for local elites to threaten NGO staff as well as local partners with violence if they did not end project work. As Dave De Vera (1996) remarked, such "everyday occurrences" required astute action. Employees encouraged nonviolent responses. On one occasion, staff took a group on a "tour" of the local prison to reduce fear of it—thereby weakening the effect of threats to "send them to prison" (De Vera 1996). They have also sought to ensure their own safety and that of partners through shrewd tactics linked to careful local action and use of political networks. Especially effective was the threat of unwelcome exposure of elite behavior through media coverage. The PAFID's unwillingness to be bullied persuaded cattle ranchers, developers, and others that "war" would not further their interests.[12]

The PAFID also ran into trouble with the New People's Army (NPA). This relationship provides another insight into how an NGO defends a "right" to work in an area despite intimidation. It also raises interesting questions about the quest for moral capital in a context shaped by *ideological* concerns. Following the split of the late 1970s when some members of the PAFID joined the NPA, the NGO subsequently followed a reformist path while maintaining some connections to the left. Indeed, one employee remarked that the organization had "good relations" even with the NPA in general (Vargas 1996). This arrangement was sensible where the NPA was active and the PAFID worked. Still, the nature of the NGO's work provided a basis for some sympathy toward the left. Indeed, PAFID analyses of the plight of indigenous people suggest overlap in what the CPP-NPA and the PAFID seek—albeit, subject to differences over violence and cooperation with "pro-capitalist" actors (Rocamora 1994; Clarke 1998; Rood 1998). That the NGO employed individuals with left-wing proclivities reinforced such congruence. Ruiz (1990: 89) captures the ambiguity here when he observes: "While both critical social movements and revolutionary movements address common concerns, they operate within a plurality of social and political spaces. In many instances, they provide mutual criticism in the spirit of critical solidarity. Here the challenge of critical social movements is fundamental, particularly in the context of the dominant statist tradition of modern politics. For these movements not only articulate a different understanding of political and ideological space, they keep these spaces open for transformation."

These comments from the late 1980s were less applicable to NGOs such as the PAFID following the split of the CPP in 1992 into factions headed by Jose Maria Sison and Armando Liwanag. The bitter feud in the CPP severely sapped the left just as the liberal-democratic regime was being consolidated. Indeed, the feud was partly over how to relate to a range of broadly leftist groups (including NGOs) independent of CPP control (Rocamora 1994; Weekley 1996).

The PAFID's relations with the NPA in Mindoro thus began to sour. A particular bone of contention was the use of GPS equipment to help indigenous communities to map CADC claims. The NPA made plain to the PAFID and the KPLN in 1994 that they opposed this practice—fearful that the technology could jeopardize their security by allowing the military to pinpoint NPA camps. They were also generally against the implementation of the CADC program. Concerned at this turn of affairs, the PAFID and the KPLN consulted with NPA commanders in November 1994 so as to share information on GPS-related work. The upshot was that the NPA would "not impede" delineation "for as long as proof is shown that it is supported by the Mangyan" (PAFID 1994b: 6). When such support was confirmed, work with the GPS resumed in 1994 and 1995. This truce was thus premised notably on the ability of the PAFID to demonstrate to the NPA that they had locally accumulated moral capital. The NPA—keen not to damage their own standing with the Mangyan—backed down once it became apparent that the NGO had strong backing from these communities.

Yet the worsening plight of the CPP-NPA—and continuing suspicion about the GPS—meant that jittery NPA commanders soon reneged on the deal. Survey work continued in 1995, with PAFID employees serving as official surveyors (DENR 1995). In December 1995, though, NPA soldiers waylaid a survey party in Occidental Mindoro, confiscating the GPS equipment, a laptop computer, and field notes pertaining to delineation. PAFID area coordinator Eric Vargas (1996) subsequently took the matter up with NPA commanders and was confronted with the "security" claim and a "ransom" demand of P500,000 (U.S. $20,000) for the return of the equipment. This episode is particularly symbolic, given the territorial significance of the technology. For the Mangyan, it was a tool to swiftly delineate "hard" territorial claims. For the NGO, it was a powerful new component in its work—and a useful means to consolidate an area-based reputation. For the NPA, the GPS equipment was feared as an inroad into its own "hard" territorial claims. Seen in this light, the NPA had nothing to gain—and potentially everything to lose—from its use.

The PAFID nonetheless shrugged off this challenge. The loss of the equipment contributed to a rethinking about its use with "security precautions to be taken at all times when using GPS equipment in field surveys"—precautions that included guarding against "confiscation by the NPA" (Romero 1996: 8). None of this, however, involved a reevaluation of the PAFID's presence in Mindoro. Work continued thereafter as before—delineation plus political advocacy to hasten CADC approval. For the PAFID in Mindoro, as with Haribon in Palawan, then, there is strong evidence of a willingness to defend territory even in the face of severe political or economic challenge. They needed political astuteness and courage of conviction. That they have done so underscores how the quest for moral capital may be linked to the pursuit of a territorial strategy. What happens, though, when territorial ambitions bring NGOs into contact and even conflict with one another?

Turf Wars

Nongovernmental organizations can work peacefully side by side. They may do different types of work and thus accumulate moral capital in parallel. Indeed, because they pursue soft territoriality, organizations may be happy to share influence.[13] There is also a possible wish for solidarity expressed through "turf etiquette." Yet "turf wars" are common, suggesting that the quest for moral capital may turn solidarity-seeking NGOs into competitors when spatial practices become territorial. This is more than a struggle over funding. As discussed, an area-based reputation can assist NGO empowerment; therefore, territorial ambitions can prompt conflict amongst NGOs.

Rules of turf etiquette seek to regulate this issue, and many NGOs have their own rules about working where other NGOs are present. Haribon's Ed Tongson (1996) professes that his organization generally "avoids stepping on turfs" so as "to avoid duplication" in light of the NGO's reputation for "pioneering work." If an NGO does work in the same area as another organization, turf may not be an issue if the NGOs have "different expertise and objectives," but, generally, to avoid the turf of others is "to respect host NGOs" (Tongson 1996). The PAFID also claims to avoid turf battles. Central here is a rule requiring a letter of invitation from a local community before getting involved. Not only is this letter a means to avoiding accusations of "meddling," but it also provides some assurance that the community is not already part of another NGO's turf (Tolentino 1996).[14]

There are also formal rules of geographical engagement propounded by NGO coalitions. CODE-NGO devised the most comprehensive set of rules, outlining procedures that members must follow in areas where "an existing NGO is operating and another NGO is planning to come in" (CODE-NGO 1995: 15). The onus is on the newcomer to check first whether another NGO is present and to ascertain its work. Where services are "already provided," the NGO must "desist from providing the same" and "instead open up new areas"—or "discuss with other NGOs already in the area other services, expertise, and resource application where it can possibly coordinate (i.e., training, capability-building programs, etc.) and cooperate" (CODE-NGO 1995: 15). Where two or more organizations work, NGOs should hold "a regular forum for dialogues" and "joint planning" to ensure "mutual respect and independence such that the work of others is acknowledged, the integrity of each agency is enhanced, its personnel developed, and conflict among agencies found in the same communities resolved" (CODE-NGO 1995: 23).

The CODE-NGO is not alone in this effort. PhilDHRRA, an NGO coalition founded in 1983—and of which the PAFID is a member—is dedicated to promoting agrarian reform and popular empowerment through "principles of partnership, between and among our members, in a network committed to strong and self-sustaining local and rural communities" (PhilDHRRA 1996: 13). As with CODE-NGO, difficulties of coordination limited success; as one member commented, "a more open/transparent relationship between and among members of the network . . . would be best for the network" (PhilDHRRA 1997: 68).

Yet these efforts have not prevented recurrent fighting. Spatial and territorial profiles overlap in complex ways that often defy regulation. Overlapping profiles are themselves subject to change as missions shift in light of new circumstances. Chapter 3 described one such shift when development NGOs such as the PAFID acquired a greener hue even as environmental NGOs such as Haribon addressed development issues. Further, some missions may be inherently spatially "restless"—as when an organization uses pilot projects around the country to disseminate new practices. To the extent that missions guide spatial and territorial strategy, shifts in the former will affect the latter.[15]

Reputation is inescapably involved here, since a territorial base conditions long-term prospects. Where and how an organization works has an important bearing on its ability to acquire moral capital with diverse partners. This is more than simply a localized accumulation of moral capital but rather involves a wider appreciation by donors and other actors of practices suggesting credibility and commitment. But if area-based action is paramount here, then the likelihood of turf disputes increases. The behavior of "competitive altruists" thus finds expression sometimes in turf wars—as the following discussion of the PAFID and Haribon suggests.

Haribon has been quite involved in turf wars—despite Tongson's assertions (noted above) to the contrary. As a pioneer in conservation, it was predictable that growing concern about the Philippine environment would mean increased competition for funds and territory. That even development NGOs began doing environmental work made the situation worse. Thus Haribon became embroiled in turf battles with the PRRM, the country's largest development NGO (Clarke 1998). The PRRM's Morales (1997) justified the move into environmental work on the grounds that there was a need for "strong political advocacy," blaming "middle-class environmentalism" of NGOs such as Haribon for having a "nature focus, not development focus." Morales thus sought to turn the PRRM into the leading Philippine NGO on the environment, with a revamped mission ("building community and habitat") and a focus on "habitat protection and management." The objective was to address "both community control over resources as well as the establishment of mechanisms for community-based resource rehabilitation and management" (PRRM 1996: 2; Dacanay 1996; Serrano 1996). Its Sustainable Rural District Development Program carried with it "an ecological component which aims to assist in building people's capabilities to undertake local environmental resource management" (Raquiza n.d.: 2). The size and *general* record of the PRRM helped it acquire funding just as funding for the area was on the increase. Indeed, funding from the NIPA scheme and the Netherlands government enabled a strong environmental project base in Palawan in the mid-1990s. This left Haribon's leaders feeling threatened by a "turf encroachment" that jeopardized their preeminence in a "traditional" stronghold (Tongson 1997).

International NGOs such as Plan International and the World

Wildlife Fund provided the biggest challenge to Haribon. The main protagonist was Plan International—with over four thousand employees and a total annual expenditure of more than $200 million in 1996, it is among the world's largest NGOs. Founded in 1937, Plan International (1996a: 23) became a global outfit specializing in humanitarian "child-focused development." It began work in the Philippines in 1961, focusing on childhood poverty in Manila—subsequently working in rural Luzon, Mindoro, and the Visayas (Plan International 1996b: 4). Problems emerged in the early 1990s as the missions of these two NGOs converged. Haribon incorporated development into its environmental focus through community-based management in coastal and upland areas—the rationale being that conservation would work only if local communities had a stake in the process. This involved paying attention to community welfare. Plan International also widened its remit. By 1996, it supplemented the traditional focus on children's sponsorship, health, and education with concerns about livelihood ("to increase food security and family disposable income") and habitat ("to ensure that children live in secure, safe and healthy habitats") (Plan International 1996b: 20). There were, too, "cross-cutting principles" for all programs, which included "equitable and sustainable access to natural resources" and "empowerment and sustainability for communities and their children" (Plan International 1996b: 20). That the two had overlapping ambitions in central and southern Luzon only added to the likelihood of turf warfare.

One flash point was Mount Isarog. Chapter 5 considered troubles surrounding the debt-for-nature swap, but what is important here is that Haribon objected to the FPE's NGO consortium for the National Park Project because Plan International was involved. Haribon opposition was based on the belief that Plan International was associated with the eviction of migrant agriculturists from another local park. Still, the consortium went ahead and Plan

International expanded its work in the area. But relations between field staff were strained in a context in which turf was unclear. One Haribon employee recalled how Plan International entered several communities in which his NGO worked and the resulting conflict as the NGOs competed for the support of the same community members. On at least one occasion, they conducted simultaneous meetings to "force" residents onto sides. Haribon's well-funded rival obtained a better attendance because it promised "tangible things" such as housing and scholarships while Haribon stressed community empowerment and capacity building. At Haribon, there was a clear dislike of this "dole out" approach, and this battle ended only when senior staff met to settle turf differences.[16]

The fiercest battle was over participation in the Northern Sierra Madre Nature Park Conservation Project in Isabela (Top 2003). Interest grew among donors, NGOs, and the government in the early 1990s as it was realized that remote Palanan encompassed the largest tract of primary lowland evergreen rainforest in the Philippines. Indeed, scientists rated it "among the most important areas for the conservation of bird diversity in the Philippines" (NIPA 1997: 7). Yet residents were among the poorest in the country and many were illegal loggers decimating the forests. As a result, the NIPA scheme identified it as one of ten "priority sites" for protection under a $7 million scheme in which a national consortium of eighteen NGOs ran individual sites. In each case one local or national NGO was selected as host. At Palanan, it was a local consortium—the Northern Sierra Madre Wilderness Foundation, comprised of people's organizations, church-based representatives, and NGOs. Conservation International (1997) was included, having been active there since 1991, when it conducted a scientific expedition, but Haribon was not included in this local consortium.

The Netherlands government thereafter became interested in developing a complementary project. Discussions were held with

Haribon, resulting in a proposal for conservation work that was the subject of a workshop in Palanan in 1995 (Teunissen 1997). A key issue was which NGO would be the lead agency. The stakes were high, since the winner would manage a multifaceted project embracing conservation, rehabilitation, and livelihood running for five years and, with a budget of $5.5 million, the largest single environmental grant in the Philippines (Ali 1997; Netherlands government 1997). Haribon employees were confident they would be picked (Haribon board minutes 1995). The NGO had extensive relevant experience and was strong in Luzon. Further, it had already received funding from the Netherlands Embassy between 1988 and 1991 for a successful marine conservation project in Zambales (Haribon Foundation 1997b; Netherlands Embassy 1997).

Yet the Dutch appointed Plan International as the lead NGO in mid-1995 in a decided snub to Haribon (Castro 1997). Even for Plan International—with an annual Philippine budget of $15 million in the mid-1990s—the award of this project was a major coup. In fact, as its worldwide annual report for 1996 proclaimed, the award was "its biggest grant ever" (Plan International 1996a: 5). In the Philippine context, the grant was a major boost in that it enabled the organization to acquire credentials in biodiversity conservation. As one employee remarked, the NGO wanted to thus build capacity because community groups were saying that "conservation was a social issue" (Ali 1997). Hence the Plan International (1996b: 22–23) Philippine report declared the project to be "part of its endeavor for environmental preservation."

The decision was a bitter blow for Haribon. The Palanan project would have extended its work in Luzon. The leadership was angry with the Dutch, suspecting them of showing favoritism toward Plan International (the latter maintained strong links to the Netherlands). It felt that the Dutch should have asked them to revise the proposal to incorporate a missing component instead of dropping them in favor of an NGO without an appropriate track record. Senior Haribon employees were also miffed that a Dutch researcher asked to review the proposal had criticized them for proposing a management structure that was what the embassy had recommended to them in the first place (Haribon board minutes 1995).

In contrast, the embassy's Hans Teunissen (1997) suggested that the decision to substitute Plan International for Haribon reflected views expressed at the 1995 workshop, where it was preferred over Haribon because it was "seen as non-political"—adding caustically that local people "seemingly preferred a non-Filipino NGO" to "a Manila-based NGO." Haribon was given the option of a minor role in the project as "technical backup" but chose instead to pullout. The project then had to be delayed as Plan International scrambled to acquire the necessary environmental expertise, even as staff found cooperation with local groups a "difficult" enterprise (Ali 1997).[17]

At Haribon, this episode illustrated what was wrong when large international NGOs such as Plan International worked in the Philippines. These organizations would "shoulder aside" Philippine counterparts in their desire to work in the country. As Haribon's Cristi Nozawa (1996) observed, there was much "resentment against foreign NGOs" for their "turfing" behavior and for the willingness of some of them to pit "indigenous NGOs against each other" to further their own ends. There was resentment, too, because of the power that some international NGOs had over the many poorly funded Philippine outfits. They could influence home country donors even as they threw their own weight around where they were donors in their own right.[18]

That Haribon and others would play the "nationalist card" is not surprising. International NGOs have a special place in the NGO sector, raising interesting questions as to the relationship between funding, moral capital, and territory. For one thing, many of them do not confront the financial constraints that Philippine NGOs face. Organizations such as Plan International or Conservation International usually come to the Philippines already armed with cash from the First World. Sometimes, though, as noted, they compete for and win funding that would otherwise go to a Philippine NGO. Either way, the problem is not a lack of cash but rather one of where to spend it. Keen to boost their own moral capital with donors and others in the First World as well as with Philippine partners, these organizations are usually anxious to play down the "wealthy foreigner" image. Indeed, they seek to get around that image by "being locally based" through work with "local government, local communities and local NGOs" (Ali 1997). The hiring of Filipinos is another way to present—literally—a "Philippine face" (Castro 1997).[19] Nonetheless, the general perception is that international NGOs are turf invaders "doing what Philippine NGOs do just as well if not better." The point is not whether that perception is "true." Indeed, international NGOs proclaim "superior" expertise in biological science and/or comparative learning from work undertaken elsewhere in the world to rebut this claim. Instead, what is of interest here is how claim and counterclaim are embedded in disputes over turf and area-based reputation.

The PAFID, too, has found itself a protagonist in turf battles. Some disputes have been with Philippine NGOs opposed to CFSA or CADC schemes. These organizations have been typically linked to the National Democratic Front (NDF)—especially hard-left factions opposed to cooperation with the state. The PAFID's ambiguous relationship with the left has been noted. For the NGO, it has entailed a low public profile when associating with leftist peers. In April 1987, for example, it attended Cordillera Day celebrations sponsored by the Communist-linked Cordillera People's Alliance (CPA) in Langawa (Ifugao)—yet the PAFID employee was told to stick to "the 'low profile' office policy . . . [and] to avoid direct involvement in politics specifically on sensitive and controversial issues" (PAFID 1987: n.p.).

Such political coyness has enabled the PAFID to accumulate moral capital with diverse partners (especially in the state and donor sectors), yet it has caused tension with some left-wing NGOs, sometimes leading to territorial squabbles. Thus the PAFID has been involved in turf-related disputes with organizations in Mindoro, a key territorial cluster for it outside of Luzon. Much of the ability of the PAFID to build a strong area-based reputation here has come down to skill at winning official land tenure instruments for Mangyan partners. Yet some NGOs and people's organizations opposed such "co-optation" by the state. Opposition centered on the Samahang Pantribu ng Mangyan sa Mindoro (SPMM)—a rival NGO-PO organization to the KPLN and the PAFID based in Occidental Mindoro. The SPMM thus attempted to halt CADC applications wherever it could, even as the KPLN and the PAFID sought to file new claims. While the latter organizations aimed to talk to the former, "little progress was achieved," such that "hardline" opponents in the SPMM succeeded in "bogging [down] PAFID's initiatives in Mindoro" (PAFID 1994b: 14; PAFID 1995a: 7).

Flashpoints in the turfing vis-à-vis international NGOs have been areas with biological diversity and indigenous inhabitants. Coron Island is a case in point. As noted, the PAFID became involved there in 1985 and established strong links. Yet the rich biodiversity has also attracted natural scientists since the 1980s. Conservational International became involved with a first contact there in 1987—soon after the PAFID's arrival. In 1992 the latter was selected by the DENR to join the pre-implementation phase of the Integrated Protected Areas System (IPAS) in Palawan. But it was encouraged to work with Victor Milan of Conservation International—something that the PAFID leadership was unwilling to do, since it believed the individual had "no heart for the [indigenous] community" (PAFID board minutes 1992). The PAFID thereafter went its own way by consolidating a role on Coron Island promoting the CADC (chapter 4). Conservation International, meanwhile, elaborated its own role in the 1990s as a proponent of the IPAS, with Palawan a focus. Indeed, Coron Island was named one of "three key sites" there to be "project priority areas"—areas in which to conduct "floral and faunal surveys" and "economic enterprise development" for local communities (Conservation International 1997: 4). It proposed to "assist indigenous communities in land tenurial rights as well as developing management capabilities for the protected areas" (Conservation International 1997: 4). This strategy featured in publicity in early 1997 even though agreement with the communities had not yet been formally obtained.

Conservation International used the occasion of the NIPAP meeting on Coron Island in late October 1996 to begin to stake a territorial claim. At that meeting, it was represented by the Philippine director Tony de la Castro and a colleague. Both were U.S. educated and raised, but Castro drew attention to the fact that he was born in a nearby town. The two made plain, too, their close identification with NIPAP—and that the key contribution of the NGO was to boost the scientific content of that program (Castro 1997; Conservation International 1997). There was a niche to be filled between the "pure research" of the university world and the "pure community organizing" of the NGO sector. His NGO would also target local business development—an area of traditional NGO weakness. Finally, Castro (1997) saw the need to provide policy support to local government. Since these were all areas in which the PAFID was "weak," he believed there was a strong need at Coron Island for the services of Conservation International (NIPAP 1997: 14). Yet the Conservation International team adopted a "softly, softly" approach when discussing these issues with islanders in recognition of both the

PAFID's local preeminence and the need to avoid open conflict that might appear unseemly. This was undoubtedly the best way in which to maximize the chances of moral capital accumulation for the organization, given that a rival was already on the scene amply supplied with local moral capital.

Despite the trappings of good turf etiquette, there was nonetheless here the outline of a serious challenge to the PAFID. For one thing, Conservation International proposed activities that would have ramifications for its own work on the island. For example, promotion of local ecotourism by the former seemed to clash with much of what the latter was counseling about "alternative development." Even more audaciously, though, the international NGO would "assist indigenous communities in land tenurial rights" and conservation management in a move that clearly challenged the PAFID's work (Conservation International 1997: 4). Conservation International also hoped to take advantage of the fact that both European Union and local government officials resented PAFID involvement in the area (Bryant 2000; Lawrence 2002). Here was a situation therefore in which one NGO's entry strategy was linked to the realization that a rival NGO's local moral capital was unevenly distributed—and potentially contestable as a result.[20]

Indeed, clashing territorial ambition was expressed in the parallel pursuit of contrasting land management models at Coron Island. As noted, the PAFID could celebrate the 1998 passage of a CADC for the island (Belen 1998). True, Conservation International supported that application at the last minute, yet the latter saw it as only a precursor to the creation of a protected area (Verian 1998). The view was that the Tagbanua could never protect the biodiversity alone. As one of those involved anonymously observed, "the so-called 'protected area since time immemorial' seems to have been inadequately guarded." Hence Conservation International support for the CADC was designed to generate enough local moral capital to advance its management model. The tactic worked: in June 1999 the island joined the EU scheme with a slim majority of residents in favor (NIPAP 1999; Lawrence 2002). The multiple designation of Coron Island under the CADC and NIPAP schemes reflects a broader process of overlapping territorial ambitions of official agencies, donors, and local community members. Here it highlights again, though, how turf disputes occur between NGOs where territorial ambitions overlap.

Thus we see that the manner in which NGOs map their missions through the promotion of a reputation for area-based work says much about how the quest for moral capital can influence an organization's decision-making procedures and indeed prospects. There are of course constraints as to where NGOs pursue their mission, since others—local bosses, insurgent armies, state agencies, donors, or people's organizations—are territorially ambitious too. Yet NGOs will often go to great lengths in order to defend territory. They become involved in turf wars. They pursue work in areas where they have a track record even when funding dries up. NGOs will defy enraged local bosses or insurgent armies. Add to this frequent harassment and/or obstructionism from local officials as well as recurrent natural catastrophes—typhoons and earthquakes, for example—and a picture emerges of tenacious NGOs keen to defend hard-won area-based reputations. Many NGOs behave in this way precisely because of the fit between an area-based territorial strategy and the ability to accumulate moral capital. They "ground" themselves by relating vision and mission to circumstances in the community as well as by looking for "spatial economies of scale" through territorial behavior. Territory thus appears to matter for NGOs because it is a key means by which organizations can build a good name.

The spatial and territorial role of NGOs is nonetheless conditioned by the political and economic dynamics examined in chapters 4 and 5. Thus, financial strategies are reflected in spatial practice. Yet the analysis in chapter 5 suggested that the choice of funding is itself linked to the quest for moral capital in complicated ways. As such, there is little scope for seeing spatial and territorial strategizing as donor determined. Similarly, NGO political strategies relate to spatial practices in a complex manner. Where NGOs work may be a spatial manifestation of patterns of critical and constructive engagement. Here again, though, the findings of chapter 4 suggest a need to avoid crude assessments—for example, that NGOs work only where state agencies wish them to operate. In short, examination of spatial and territorial strategizing underpins a view of NGO behavior as too complex to be pigeonholed through narrow explanation.

The preceding three chapters have illustrated the utility of thinking about NGOs as organizations involved in a multifaceted pursuit of moral capital. They are seemingly keen to build reputations as moral and altruistic agents resorting to varied political, financial, and territorial strategies in order to do so. There are many dangers in such a quest, not least of which is the risk that excessive zeal in pursuing moral capital may put off partners, thereby paradoxically reducing the stock of capital. In some cases, acquiring moral capital in one quarter may result in the loss of moral capital in another quarter as an NGO becomes embroiled in a difficult and potentially damaging zero-sum situation. What this suggests is that the quest for moral capital demands of an NGO great skill—mixed with a measure of luck—in dealing with a wide array of people as well as situations often not of its own choosing.

It is also important to remember that NGOs are certainly not animated exclusively by moral capital calculations. Other social, political, and economic factors condition action such that the sources of NGO empowerment and decision-making can never be captured by one-dimensional explanations—something readily acknowledged in this book. Still, the preceding analysis has sought to shed new light on a hitherto neglected aspect of the subject. It has done so by assessing the implications of NGO efforts to capitalize on moral standing in order to acquire the power to effect change. Organizations are thus involved in decision-making whereby lofty aims are translated through political, financial, and territorial strategizing (and vice versa) into a desired ability to "do good." Yet this ability is invariably linked to the acquisition of moral capital or what the PAFID aptly called its "gold mine." Here, then, are grounds for describing NGOs as moral entrepreneurs and consummate strategists.

CHAPTER 7

Conclusion

Morality Plays

Much behavior by the NGOs examined in this book may be summarized in two words: morality plays. There are various meanings, but let me highlight two. The first is concerned with dramaturgy, as NGOs give a finely tuned performance on environment and development matters, "playing" to an audience of partners. To generate an image for doing "good" through a story spun from words and deeds is to attempt to build reputation. The second meaning reveals what morality can "do" for NGOs—it is about a particular sort of strategizing. Here, NGOs would link their names to moral issues or "causes," not only out of individual and collective belief but also sensing that to do so may be a key means of organizational empowerment. To the extent that morality "plays well" in society, NGOs may "play up" their association with moral principles and practices, albeit not in a strident or excessive manner, for fear of alienating partners. Indeed, skill is required in getting the balance right between "boosterism" and modesty in the quest to accumulate moral capital. Here is a notion of an organization seeking "to play one's cards well . . . to use one's resources in the most effective manner" (Guralnik 1986: 1092).

I have explored in this book the logic and implications of a type of strategic thinking that underpins a multifaceted NGO quest for moral capital. The term *moral entrepreneur* has been used here, and with reference to two Philippine NGOs working in the environment and development field. To what extent, though, is it fair to suggest that this understanding has explanatory purchase on NGO behavior generally? And what are the implications of our argument for a broader understanding of NGOs as reputation-seeking actors operating in a world of moral capital production and consumption?

Seeing NGOs as Moral Entrepreneurs

In assessing the broader utility of the approach elaborated in this study, it is important to reiterate what this book does and does not argue. It does not claim, as neoclassical theorists such as Gary Becker seem to do, that human action can be reduced in all of its complexity to mere economic calculation. However, the book *does* suggest that NGO empowerment can be *partly* a matter of an instrumental form of strategic rationality understood as "consistency over action in line with preferences" (Bridge 2001: 209). It can be said to occur when individuals or groups "choose the best available means to achieve what they understand to be in their interest," even as that understanding can shift over time as ends sometimes adapt to new circumstances (Chong 2000: 12). In the case of the NGOs studied, I suggested that the "best available means" relate to being culturally resourceful as NGOs promote a (shifting) role for themselves in environment or development work. Economic factors can be important, but social and political factors can also be central. The quest for moral capital is, then, linked to an instrumental form of strategic reasoning but in a way not synonymous with economic logic.

The quest for moral capital involved the Haribon Foundation and the PAFID in diverse, multifaceted *political* strategies involving notably critical engagement with state agencies and constructive engagement with local communities. Interestingly, these strategies did not always bear fruit as might have been anticipated. There was evidence that political strategizing in aid of reputation building did not always lead to economically positive outcomes. As the Haribon experience showed, an organization might even lose funding as a result of political advocacy. Still, while funding and sometimes even moral capital was lost in some quarters, support was acquired elsewhere. In addition, NGOs did not always attain formal political goals. The PAFID discovered this, for example, in its battle to halt "invasive" economic activity in Tagbanua lands.

Yet the resort to political strategy was nonetheless worthwhile. As moral capital was thereby accumulated, notably with donors, state agencies, and local communities, both NGOs were better able to press for change in an incremental and intertwined process of reputation building and political outcome. For example, greater public awareness about the "evils" of logging plus the development of a politically assertive indigenous community were fruits of the labor of Haribon and the PAFID, as well as being reflective of their enhanced social standing among an array of partners. Yet these achievements served, in turn, to ratchet up expectations for change in these sectors, providing further opportunity for these and other organizations to promote organizational goals. Indeed, both NGOs appeared to make modest if uneven headway in the promotion of their reform-minded visions and missions even as they were able to consolidate their own roles in a wider process of social reform.

The pursuit of moral capital also left its mark on the financial strategies that the PAFID and Haribon devised in looking for funding. The need for funding is a well-known source of anxiety and infighting. Yet that need—and how organizations go about satisfying it—seems intimately associated with the effort to have a good name. As chapter 5 pointed out, this has important implications for NGO action. There was, for example, Haribon's "disproportionate" effort to promote a diverse range of income streams. Most of these initiatives were quite minor financially, but they held out the prospect of enhanced autonomy—hence the effort to boost membership, sponsor concerts and films, or peddle merchandise to a middle-class clientele. Each of these efforts also provided benefits to the organization in terms of enhanced name recognition and respect for the NGO cause. In addition, "insightful agility" led both organizations to aim for a diversified donor portfolio—national and international groups, as well as private and public organizations—with particular attention devoted to donors such as the MacArthur Foundation or Misereor, who afforded maximum flexibility. Finally, there was the rejection of funds: "tainted" business money by Haribon and the pullout from the Low-income Upland Communities Project (LIUCP) by the PAFID were our examples.

These various stratagems make sense when it is recognized that NGOs seem to devise financial strategies with an eye to the simultaneous accumulation of funding and moral capital. For the PAFID, for instance, the LIUCP pullout was a major financial blow that was nonetheless acceptable since the alternative was seen to be worse—alienation from Mangyan partners. The latter would have had disastrous consequences for the NGO's standing as a leading national defender of indigenous people as well as for its future work. This sort of event is an example of strategic rationality in action reflective of much considered thought. Even as the PAFID left the LIUCP in solidarity with the Mangyan it went to great lengths to justify its actions to DENR officials—and to reassure them of the NGO's continuing desire to work with that agency.

Finally, chapter 6 described how moral capital becomes entangled in the spatial and territorial strategies of NGOs. For both the PAFID and Haribon there was elaboration of soft territoriality, with territorial clusters created in various parts of the country. The generation of multiple area-based reputations—Haribon in parts of Luzon and Palawan, the PAFID in central and northern Luzon, Mindoro, Palawan, and southern Mindanao—has been a complex outcome of missions, personal contacts (social capital), and sheer hard work. The importance of these area-based reputations was seen in the determination of each organization to continue to work in these regions even when political and economic events turned against them. Such was the case for Haribon at Mount Isarog when funding dried up in the aftermath of the debt-for-nature swap debacle. Such was the case for the PAFID in Oriental Mindoro when, first, funding collapsed in the wake of the LIUCP pullout and then the New People's Army confronted it over its GPS-aided mapping program.

The importance of area-based reputations was also highlighted when these NGOs became embroiled in turf wars with other NGOs. This was the case as Haribon confronted Plan International in Luzon and when the PAFID crossed paths with Conservation International in Palawan. These "wars" are often soon resolved, but tension sometimes lingers long thereafter. It is the fact that they break out at all that is of interest—and that they are often partly linked to territorial strategizing and moral capital accumulation. Here, there was evidence that the quest for moral capital is not always compatible with NGO solidarity, rules on turf etiquette notwithstanding. The instrumental form of strategic rationality that underpins the pursuit of moral capital can turn even solidarity-seeking NGOs against one another from time to time.

The preceding inquiry has suggested that the two case-study NGOs have been generally adept moral entrepreneurs. They have been able to elaborate complex political, financial, and territorial strategies in aid of their own empowerment through a multifaceted quest for moral capital. True, things have not always gone their way. Indeed, Haribon suffered a decline in reputation in some quarters following the anti-logging campaign as reservations were voiced about "overly political" action by it. The moral entrepreneur does not win every political, financial, or territorial battle. There are bound to be occasional defeats—and an associated loss of moral capital—over the years. It is worth remembering that reputation is invariably contingent, since it depends on perceptions that an NGO can never control. Moral capital is "always vulnerable to perceptions of serious betrayal of, or incapacity to pursue, valued goals and principles" (Kane 2001: 41). In this light, the moral entrepreneur needs to be prudent in thought and deed. There is need for a complex approach encompassing diverse partners and with an eye to promoting an ability to work in keeping with a chosen vision and mission—a choice moreover that may shift over time. Haribon no longer has a preeminent reputation deriving from its status as a "pioneer" in Philippine environmentalism, but this has perhaps as much to do with the increase in the number of NGOs in that sector than it has to do with the vicissitudes of the Haribon reputation per se. That NGO is, after all, still going strong today with a mission and project profile not unlike the one of a decade earlier.[1]

Moral Capital Sells

It might be argued that these two NGOs ideally fit the moral capital perspective. That perspective may help in understanding a small set of reform-minded "national" NGOs, even though the assumptions and elements of moral capital may not permit the perspective's application to other sorts of NGOs, let alone other types of actors. As the introduction noted, this challenge necessitates further empirical work. Indeed, the case study method used positively invites such work. New conditions might include different national and/or cultural contexts, different political regimes, different sizes or types of NGOs, or different NGO ideological proclivities. I do not wish to second-guess what these sorts of inquiry might reveal. Perhaps, though, some carefully targeted speculation is in order here.

Drawing possible insight from my research (as well as that of others) in the Philippines, how might the moral capital approach fair, then, if applied to other sorts of NGOs?

First, how might the perspective stand up if applied to the case of politically radical Philippine NGOs? I am thinking of left-wing NGOs that developed during and after the 1970s, forming part of the CPP-linked National Democratic Front (NDF) as well as kindred organizations opposed to armed struggle, such as the social democrats and democratic socialists. The general picture to emerge here is of a group of organizations divided over ideology as per shifts in the CPP itself (Rocamora 1994; Boudreau 1996; Clarke 1998). These schisms aside, how does the leftist outlook shared by these NGOs affect strategic sensibilities about moral capital?

The prominence of ideological debates and associated politics ought not to blind us to the *possibility* that the operations of left-wing NGOs are not that different from those of reformist NGOs. Indeed, there are grounds for suspecting that the former are every bit as embroiled in a quest for moral capital as are the latter. Thus they need to attend to their reputations even as they compete with each other and reformist NGOs for funding. In this regard, Clarke (1998: 124) observes that "a history of effective service delivery" was important because it enhanced "their 'legitimacy' in the eyes of other political actors." True, some have refused to seek funds from "ideologically suspect" outfits such as USAID. Such a stance is consonant with the pursuit of ideological consistency. Yet it also helps to boost the credibility of these organizations with key partners—"radical" community groups or elements in the CPP-NPA, for example. Such behavior is reminiscent of that of reformist NGOs such as Haribon when they refuse funding from "dirty" business in order to safeguard reputations with reform-minded partners. In short, the question of ideology seems to be important not because it means that the logic of moral capital is suspended but because it conditions the choices that NGOs make in terms of partners—that is, from *whom* they seek moral capital.

Second, how might the moral capital approach fare if applied to the case of either smaller or larger NGOs than those featured in this book? Let me begin with the case of small local NGOs, of which there are many in the Philippines. It is possible that the "pool" of partners is smaller, in keeping with the reduced scope and geographical area of operations. Some might think at first glance that this would put a crimp in a multifaceted quest for moral capital. Indeed, there is frequent complaint by small local NGOs that they cannot win the attention of key people in the media, state agencies, or donor organizations in a way that national or international NGOs can do. Yet small, local NGOs are no less involved in the quest for moral capital than national or international peers. In Cebu, for instance, an NGO called the Soil and Water Conservation Foundation (SWCF) built a name for doing technically proficient work in upland watersheds. The contacts of American-born founder Bill Granert were certainly helpful. Moreover, this organization earned kudos with an array of actors, including local government, the regional DENR office, local and national NGOs (including the Environmental Legal Assistance Center and Tambuyog), and donors such as the Australian government and the Ford Foundation. This multifaceted accumulation of moral capital can be largely attributed to the painstaking work of this organization in watershed conservation and its willingness to take on powerful opponents when necessary. In the mid-1990s, for example, the SWCF became embroiled in conflict with the Ayala Land Corporation—a key holding of the powerful corporate empire owned by the Zobel de Ayala clan—over a proposal by the latter to build a golf course in an endangered watershed.[2]

The experience of the SWCF may represent that of other small NGOs in the Philippines, suggesting that it is *plausible* to assume that this sort of NGO might behave strategically in the quest for moral capital. Even as political and economic developments have widened the potential pool of partners, the partial devolution of power to local government after 1992 enhanced the visibility of local NGOs in official thinking (Clarke 1998; Silliman and Noble 1998). In addition, donor agencies also sought to "reach out" to these NGOs and people's organizations through "equitable" and "targeted" assistance. Still, it does not necessarily follow that reputation-building endeavors are the same for small local NGOs as for larger national or transnational NGOs. Much appears to depend on factors such as geographical remoteness and the life histories of NGO employees. One could imagine how isolated NGOs might be especially reliant on what local people think of them and their work. As with the tightly knit communities of southern Europe described by the anthropologist F. G. Bailey (1971a) and others in the early 1970s, individual and group reputations might be even more important for regulating social conduct and opportunity in these organizations than is true for NGOs with a wider spatial and territorial reach. It could be plausibly assumed that the social forces of localism might be an especially powerful inducement—if one were needed—for NGOs to assiduously accumulate moral capital.[3]

Finally, let me turn to the case of the large international NGOs that operate in the Philippines—several of which have appeared in these pages already as antagonists of the PAFID and Haribon. In some respects, the situation of these international NGOs is the exact opposite of that of the small local NGOs just considered. Thus community work for these giants is often simply the empirical application of social or environmental models developed elsewhere. For example, there is the biodiversity hotspot model of Conservation International (1997) and the foster parent model of Plan International (1996a). As a result of this sort of approach, it is argued, these organizations never acquire roots locally in the way that local or even national NGOs seem to do.[4] However appealing this argument is to Philippine NGOs affronted by the wealth of international counterparts, it is probably not widely applicable in practice. As chapter 6 noted, international NGOs like Plan International or the World Wildlife Fund have been working in the country longer than many local NGOs. Hence they sometimes earn enviable *local* reputations, as with Plan International in Cebu, for example, where work on water installation since 1979 has earned it praise in the community (Van Engelen 1997).

What is more, the pursuit of moral capital would appear to be for them much more than simply a matter of earning kudos locally. Indeed, the quest for moral capital seems to occur simultaneously at multiple scales and with reference to a heterogeneous group of actors around the world. For NGOs such as Conservation International, WWF, or Plan International, that group would appear to include state agencies (in host and home countries), the general public (especially in the United States and Europe, where it is targeted for sponsorship), private donors, fellow NGOs, the media, United Nations officials, and so on. The international character of these NGOs is often something that employees are proud of—as one suggested, it gives them a "wider perspective" than that of local or national NGOs (Ali 1997). What is important here is that this perspective almost inevitably entails an acute strategic sensibility as the organization seeks to build a reputation simultaneously in many places with many partners. This sensibility thoroughly permeates the reports of these organizations. For example, Conservation International (1996: 2–4) emphasizes its standing as a world leader in the scientific community even as it boasts that "decision-makers in many countries have begun to rely on CI's conservation strategies."[5]

Here too, though, it does not follow that because international NGOs may be moral entrepreneurs like national or local Philippine NGOs, their specific practices adhere to the same pattern. It might be plausible to assume that the *international* character of the NGOs means that these organizations attach *relatively* less importance to the generation of a good name in any given location than do their locally based counterparts (Kellow 2000). Such a view would need to be substantiated through careful empirical work that unpacks the political, financial, and territorial strategies of these NGOs in relation to moral capital. What is clear from this all too brief discussion is that the moral capital approach would seem to hold promise as a *possible* explanatory framework for various sorts of NGO—and hence, not just the type of reformist national NGO featured in this book. Indeed, further empirical work may even shed new light on "traditional" issues of ideology and scale in NGO action—highlighting how these issues, for instance, mediate or refract the reputation-building endeavors of NGOs.

Finally, a brief word is useful here on moral capital and how it might be relevant to the understanding of *other types of actors*, such as businesses or politicians. With regard to the latter, Kane (2001) has already offered an insightful account of how moral capital can play a crucial part in building political fortunes. Using case studies of such political leaders as Charles de Gaulle, Aung San Suu Kyi, and Nelson Mandela, he argues that moral capital is important because it helps to mobilize political support, create new strategic opportunities, and enhance political legitimacy. Contrary to the view of the political arena as being essentially amoral, then, Kane asserts that moral judgments can be an important resource in their own right in the vicissitudes of political life. It would seem that the fact that many do not expect political leaders to have moral standing today may amplify the benefits that accrue to those leaders who are the subject of positive moral judgment.

If moral capital has purchase on "amoral" politics, it is also *plausible* that it *might* have some role to play in the business world. Clearly great care is needed here to distinguish between sophisticated PR and soothing moral rhetoric, on the one hand, from a process whereby the strategies and opportunities of firms are conditioned in part by the moral judgments of others (Richter 2001). The effect of moral capital on a business organization's life may be great in those sectors in which competitive advantage is seen to go hand in hand with ethical appeals of various sorts—"fair trade" firms, organic food producers, and so on (Kaplan 1995; Goodman and Goodman 2001; Hughes 2001; Bryant and Goodman 2004). Indeed, the possibility of reaping competitive advantage at the level of the individual firm may be even greater still in those sectors or industries not hitherto associated with positive moral judgments, such as natural resource extraction industries, pharmaceutical companies, and so on (Prakash 2000). This is so because the unexpected nature of "good conduct" in a "bad sector" might result in especially warm plaudits from NGOs, regulatory agencies, and consumer groups for "exceptional" behavior. That said, claims here must be critically assessed for "hidden transcripts" (J. Scott 1990). Not only do competitive pressures suggest limited scope for potentially costly ethical "concessions" but ubiquitous appeals to "corporate social responsibility" render it ever more difficult to overcome the mounting suspicion that there is little substance to such reputation building (Richter 2001).

The Social Production and Importance of Moral Capital

Our inquiry has taught us that moral capital can matter enormously for NGOs. Because it can contribute to their empowerment, organizations such as the PAFID or Haribon devote effort to political, financial, and territorial strategizing that aims to promote the good name of the organization. For this reason, reputations should matter too to all that seek to understand the meaning and significance of NGO strategic behavior.

Much remains to be discovered, though, in attempting to understand the social dynamics surrounding the quest for moral capital. While attention here has remained focused on what NGOs may do in order to acquire moral capital, very little has been said about the *production* side of the equation. Work needs to be done on how reputation producers create moral capital for an NGO. Since reputation is a social construction (albeit one usually linked to tangible practices), there is a need to know more about how specific individuals or groups—acting alone or in conjunction with others—come to produce a reputation. How is this production process related to the particular needs, interests, and views of reputation producers? I have observed that NGOs attempt to acquire moral capital in part by conditioning these needs, interests, and views. That is not to say, though, that producers do not have their own agendas—think of the preoccupation of many donors since the early 1990s with the "professional development" of the NGO sector, for example. In any event, it is probably safe to assume that reputation producers are not "disinterested" observers. Indeed, reputations would appear in part to be "grounded in the needs and the perspectives of those who put forward claims about those reputations" (G. Fine 2001: 20).

Yet the determination of reputations would appear to reflect an unequal contribution on the part of producers. While each producer is involved in the creation of a reputation for an NGO partner, some are influenced greatly by what others have to say about the organization. Here, the role of "reputation entrepreneurs" requires close attention (G. Fine 2001). These producers may have a strong stake in the reputation of an NGO for various political, economic, or cultural reasons. They may simply have a particular gift for providing a compelling interpretation. Whatever the reason, the role of these reputation entrepreneurs—who may be close allies of the NGO, political opponents, donor representatives, or perhaps others with an ax to grind—would seem to be salient in any account of the production of moral capital. This would appear to be the case especially when an NGO has been around awhile—that is, when it acquires a distinctive and discernible *history*. It is easy to forget when studying NGOs, especially in the Third World, that these organizations are beginning to develop historical profiles not unlike those of some of their First World counterparts (Wapner 1996; Keck and Sikkink 1998). In the process, it becomes important for both NGOs and those who would study them to recognize that history is often littered with good reputations turned sour as reputation entrepreneurs succeed in making the case for less complimentary interpretations (Moorhead 1998). What is more, it is not just "great" cultural or political figures that need to be fearful for the place allotted to them in history (Becker 1992; Lang and Lang 1990).

There is, too, in all of this the question of the broader social role and utility of NGO reputations. This question certainly congers up a Durkheimian perspective in which reputations are seen to be about social integration through "symbolic work." Leaving aside the narrowly functionalist view of that perspective, though, there are still ample grounds for investigating the macro-level implications of the micro-level strategizing examined in this book. What is the overall significance to Philippine society (and other societies besides) of a situation in which NGO behavior and empowerment is partly conditioned by the quest for moral capital? Furthermore, how might moral capital fare as an exploratory concept in settings characterized by sharply different cultural contexts from those featured here? Our concern in this book has been to focus on the strategic thinking of two "Westernized" Philippine NGOs as well as the internationally shaped circuits of moral capital that are associated with them. It may be that investigations of other NGOs in contexts less shaped by international (i.e., Western) cultural norms and values may lead to exciting alternative takes on strategy and rationality in relation to moral capital, including of course the very definition of "moral capital" itself (Howell 1997).

These lines of analysis surrounding the social production and importance of moral capital would undoubtedly raise interesting new angles on the subject matter at the heart of this book. I suspect what they would not do, however, is to cast doubt on the important role played by reputation in the affairs of NGOs. These organizations, it seems to me, are quite correct in seeking to cultivate reputation through strategic action. To do otherwise might be to abandon the field to reputation entrepreneurs holding a less "charitable" view of them. It would also be to forgo a chance to actively promote their own empowerment—an outcome without which even the best-intentioned vision and mission will come to naught in the end. With so much at stake, this is one risk that most NGOs dare not take.

Notes

Chapter 2. The Quest for Moral Capital

1. The focus of this study is on the more instrumental aspects of strategic rationality associated with two leading internationally connected NGOs in the Philippines. A discussion of a wider (i.e., noninstrumental) practical rationality is therefore beyond our purview here (cf. Espeland 1998). It is important to stress, though, that rational strategizing discussed herein is only one part of a larger picture when discussing human rationality.

Chapter 4. Political Virtuosity

1. The intervention by this guest—a member of a prominent NGO—is revealing on two counts. The individual firstly succeeded in antagonizing the audience with a paternalistic demeanor. Indeed, his intervention strengthened opposition to the NIPAP because residents feared that such disrespectful conduct might be part and parcel of the scheme if it was implemented. Secondly, and in the course of his intervention, the guest sought to curry favor with the Tagbanua by identifying his NGO with the "good work" of the PAFID on the island. Here we have a case in which one NGO seeks to boost its reputation locally by publicly associating with another NGO already in good repute in the area. The attempt failed, judging by the reaction of the assembled Tagbanua.

Chapter 5. Financing Prophets

1. The reference here is to the anti-Spanish Katipunan Revolution of 1896 in which a small band of insurgents was able to challenge a powerful colonial force due to the rapid spread of rural popular support for the rebels.

2. There is an irony here, given Hollywood's bad environmental reputation in Southeast Asia arising from the production of films such as *Apocalypse Now* (filmed in the Philippines) and *The Beach* (filmed in Thailand).

3. Anonymous interviews. Much depends on highly specific labor supply and demand situations that place even NGOs in competition with one another. Haribon, for instance, found in the mid-1990s that it was losing trained community organizers to international NGOs moving into the Philippines. Salaries were tripled in some cases. This practice has been deeply resented by Philippine NGOs.

4. Anonymous interviews. True, NGO work can also be a useful apprenticeship in environment and development work. Skills are learned that may enable individuals to later move on to more financially lucrative careers.

5. This paragraph is based on numerous interviews as well as direct observation in both Manila and elsewhere in the country during 1996–1997.

6. In contrast to the Ford Foundation or the MacArthur Foundation, Philippine equivalents like the Ayala Foundation or the San Miguel Foundation are among the least important of funding partners. International NGOs provide funding but face their own financial uncertainties. Indeed, they sometimes compete with local NGOs for project money. Official ODA providers such as USAID and CIDA are hemmed in by budgetary pressures, staff cutbacks, and demands for "value-for-money" at home.

7. The concern here is with Philippine businesses that provide donations directly or indirectly (i.e., foundations). I am not concerned with international private foundations of a business pedigree such as the MacArthur Foundation or the Ford Foundation, which, as noted, enjoy a reputation in the Philippines as "blue-chip" donors.

8. That business leaders sit on the board helps the NGO access business networks. In some cases, board members directly underwrite operations in times of financial dearth. Belen King has done so, as has Kalaw (anonymous interviews).

9. The board considered creating a dirty list in June 1993 but was divided on the matter (Haribon board minutes 1993).

10. Another example concerns the Philippine mining giant the Benguet Corporation. In 1997, it was on the dirty list even though it had been a corporate

member of the NGO as late as 1987 (Haribon Foundation 1987; Holopainen 1997). There is no evidence to indicate that environmental performance had changed over the period.

11. There is, too, rivalry between NGOs of a comparable size such as PRRM and Haribon. This paragraph draws on an array of anonymous interviews.

12. In the case of another NGO-managed funding mechanism, NGOs for Integrated Protected Areas, Inc. (NIPA), a consortium of eighteen NGO and PO networks, was formed in December 1993 to protect biodiversity hotspots. The NIPA was formed, though, only after NGOs protested to the donor (the World Bank) that it was inappropriate to choose two NGOs (WWF and PRRM) to run the whole project. Difficulties in project implementation resulted in further NGO tensions (NIPA 1996).

13. Anonymous interviews. Haribon's stake declined such that by 1995 funding was P75,000, of which the consortium received nearly P60,000. Haribon opposition to Plan International was part of a wider battle over turf discussed in the next chapter.

14. The creation of the Foundation for a Sustainable Society (FSSI) in 1995 through a debt-for-development swap by the Swiss and Philippine governments was a further opportunity here. Based on an endowment of $17.3 million, the FSSI targets "sustainable production" in the country. Here again, though, the need to have at least a two-year track record in the sector tends to hinder the chances of new organizations (CODE-NGO et al. 1995).

Chapter 6. Mapping the Mission

1. Attention to NGO territoriality raises the point that there are few truly "national" Philippine NGOs. As we have seen already, the case NGOs derive support of various kinds from "global" actors, including large NGOs (such as the WWF) and donors. They are also embedded to a greater or lesser extent in regional and global networks of activism, thereby suggesting complex multi-scale fields of action and territorial engagement—through, for example, "grassroots globalization networks" (Routledge 2003; see also Keck and Sikkink 1998). For our purposes, though, the focus is on territorial strategies by the PAFID and Haribon concerning the Philippine context alone—which is indeed most of such strategizing by these NGOs.

2. Cases also exist—for example at Didipio, Neuva Viscaya—where the PAFID assists displaced groups to make claims to *new* lands in which they reside (direct observation, October 1996).

3. Staff field reports bear testimony to the lifestyle. A common problem is

prolonged traveling due to floods, landslides, typhoons, or vehicle breakdown. This was certainly true on one trip I made with staff in 1996. We arrived at our destination in the middle of the night—many hours later than planned—due to floods and vehicle problems.

4. Rice's networks built up over decades were especially important. There was a religious network spanning the Protestant and Catholic faiths. That Rice was also an anthropologist meant that he had a fine network of community-level contacts via researchers. Thus links to Palawan were facilitated through his knowing the renowned American anthropologist James Eder.

5. Julian De Vera was a long-standing supporter and board member of the PAFID. He was also the father of Dave De Vera (executive director in the 1990s) and Prospero De Vera (board member in the 1990s). For an account of life among the Mangyan, see Conklin (1957).

6. One example of the PAFID's remaining in a community after an initial project was completed is that of Sitio Bailan in Barangay Binli. Here, a CFSA was first arranged in the 1980s. This was followed by a livelihood project funded by the Canadian Hunger Foundation (PAFID n.d.).

7. These processes are linked. State devolution offers new opportunities at the local level for "national" NGOs even as decentralization by NGOs increases the pool of experienced NGOs working in communities, thereby facilitating state devolution.

8. A particular source of concern was Haribon Palawan. In the mid-1990s, there was a rumor about possible collusion between one employee and a Taiwanese investor hoping to buy land from an indigenous community. Following an investigation by Manila, that employee was told to sever links to the investor or leave Haribon. This sort of embarrassing incident—potentially damaging to the Haribon name—led Manila to assert tighter control in Palawan via donors.

9. It was perhaps with this episode in mind that the following item in the report of the 1992 PAFID staff retreat was composed: "Cultural sensitivity was underscored both in dealing with client-communities the PAFID is serving and to other staff members who come across different cultures, with different styles, methods of thinking, and with varied work orientations" (PAFID 1993c: n.p.).

10. The DENR introduced a moratorium on commercial logging in Palawan in 1992 and canceled the remaining timber agreements there in 1994. Yet, as Arquiza (1996: 143–44) notes, the "need for vigilance" remained as Alvarez and others sought to circumvent the DENR order.

11. Recall that the decision to stay in Mindoro was made *prior* to the award of new funding—and hence was not conditional on receipt of new funds.

12. PAFID leaders believe that its reputation ("gold mine") requires "cour-

age to penetrate high-risk areas" (PAFID 1994c: 12). Intimidation persisted: in 1994, timber plantation owners threatened the PAFID because they feared their rights would be invalid if indigenous people won ancestral domain (PAFID board minutes 1994).

13. Hard territoriality is about asserting *control* while soft territoriality is about asserting *influence*. Hence, the clash of actors in pursuit of the former is usually more severe than the clash of actors in pursuit of the latter. Such a case is the struggle between the Philippine government and the CPP-NPA.

14. For one letter, see Residents of Mapayao (Nueva Vizcaya) (1992). Such a letter does not guarantee the absence of another NGO in the area but does tend to reduce if not eliminate turfing.

15. Are spatial and territorial strategies guided by other factors? There is evidence that ideology has been a factor, but its declining political saliency in the 1990s may have reduced overall impact on NGOs—especially those of interest here. Another possibility is personality, as evidenced in the differences between Kalaw and the PRRM's Horacio Morales, which are said to have heightened tension between the NGOs. Yet these personal likes and dislikes scarcely determine overall spatial and territorial thrust. Finally, there is evidence that NGOs will sometimes go where donors wish them to be, resulting in "mission creep." Yet, this explanation is not satisfactory. Donor priorities are themselves often shaped by NGOs—for example, the rush of donors to support work in Palawan *after* Haribon and others demonstrated the need for action. The spatial and territorial strategies of NGOs also usually reflect complex responses to the quest for moral capital in which economic *and* noneconomic factors are important.

16. Anonymous interviews.

17. The relationship between the NIPA and Dutch projects was tense. There was mutual suspicion over turf, and NIPA feared that the "bountiful" cash associated with the Dutch project would "corrupt" local people, diluting the conservation focus (Castro 1997; Isberto 1997). Plan International, meanwhile, certainly had strong Netherlands links. A 1996 report indicates that four of the twenty-one international board of directors were from the Netherlands—more than any other country. The Netherlands was also the leading donor country to Plan International—in 1996 contributing a whopping 44.8 percent of the total—compared with the next largest contributor, Japan, providing 13.7 percent (Plan International 1996a: 1, 20).

18. The WWF is often cited as an example of an international NGO prepared to bully Philippine NGOs. The equivalent example of a "bully" cited from among Philippine NGOs is PRRM. For example, one veteran NGO leader anonymously described the latter using the words "pera rami"—a play on the

pronunciation of PRRM, in Tagalog meaning "more money." For an excellent account of the PRRM, see Clarke (1998: 138–64).

19. The Philippine reports of these NGOs also emphasize a "local" ethos. Thus Conservation International (1997: 1) has a "strong emphasis on local capacity building, close coordination and partnership with in-country institutions." Plan International (1996b: 19) proudly mentions *local* recognition for its work—namely, the award to Plan Cagayan of the "Fighting Cock Outstanding Regional Award" by the local Progressive Alliance of Citizens for Democracy.

20. A ham-fisted display of territorial ambition occurred at the NIPAP meeting when another NGO—the International Marine Alliance (IMA)—sought favor with the Tagbanua. The IMA's coordinator repeatedly identified both the IMA and Conservation International with the PAFID as lead NGOs that would implement NIPAP. The PAFID's De Vera reacted angrily to this ploy to insinuate the IMA into local people's affections by tapping into his NGO's local moral capital: "how dare he identify PAFID with the IMA or with himself!" (De Vera 1996).

Chapter 7. Conclusion

1. In July 2001, for example, the Haribon website made explicit this broad continuity of aims and objectives: "It now seems very appropriate that the foundation was named Haribon, coined from the words 'Hari' and 'Ibon' which loosely translates to 'King of Birds' since the foundation evolved from a bird-watching society to a natural resources conservation foundation. And one of Haribon's major projects are research and publications on biodiversity using birds, and the Philippine Eagle in particular, as the flagship species since birds are a good indication of the state of an ecosystem and the overall condition of the environment." See www.haribon.org.ph.

2. This summary is based on interviews and direct observation while on a field trip to Cebu City in May 1997. I met with representatives from the SWCF, other NGOs, and the local DENR. Additional interviews with individuals familiar with this NGO were also conducted in Manila.

3. Of course, even many remote Philippine communities are not as isolated from the outside world today as they were. Given this interconnectedness, one might expect that the quest for moral capital would become complex, inasmuch as it would relate increasingly to both local and nonlocal partners.

4. This particular critique is a distillation of the sentiments expressed by various representatives of Philippines NGOs in the course of personal interviews conducted during 1996 and 1997.

5. The sheer spatial and organizational reach of these NGOs can be breathtaking. Consider the case of Plan International, which in the mid-1990s worked in forty developing countries (as well as in developed countries) with 800,000 foster parents alone. To this must be added the many thousands of community organizations and families in question as well as local, national, and international governance (Plan International 1996).

Bibliography

Printed Sources

Adbusters. (2001) "Toxic Culture." Special issue of *Adbusters: Journal of the Mental Environment* 36 (July–August).

Alagappa, Muthiah, ed. (1995a) *Political Legitimacy in Southeast Asia*. Stanford: Stanford University Press.

——. (1995b) "Introduction." In *Political Legitimacy in Southeast Asia*, ed. Muthiah Alagappa, 1–8. Stanford: Stanford University Press.

Albano-Vitug, Marites. (1996) "Isang Kuwento ng Pakikibaka sa Bolinao." *Haribon Quarterly 2, no. 4* (October–December): 3–4, 9.

Aldaba, Fernando T. (1992) "The CODE-NGO: Unifying the Development NGO Community." *Development NGO Journal* 1, no. 1: 1–14.

Alegre, Alan, ed. (1996a) *Trends and Traditions, Challenges and Choices.* Manila: Ateneo Center for Social Policy and Public Affairs.

——. (1996b) "NGO Relations with Churches." In *Trends and Traditions, Challenges and Choices,* ed. Alan Alegre, 95–101. Manila: Ateneo Center for Social Policy and Public Affairs.

——. (1996c) "The Rise of Philippine NGOs as Social Movement: A Preliminary Historical Sketch, 1965–1995." In *Trends and Traditions, Challenges and Choices,* ed. Alan Alegre, 2–48. Manila: Ateneo Center for Social Policy and Public Affairs.

——. (1997) "Singing Your Funding Song to Donors?" *Philippine NGO Memo* 1: 12–13.

Allahyari, Rebecca Anne. (2000) *Visions of Charity: Volunteer Workers and Moral Community.* Berkeley: University of California Press.

Alvarez, Sonia, Evelina Dagnino, and Arturo Escobar, eds. (1998) *Cultures of Politics, Politics of Culture: Re-visioning Latin American Social Movements.* Boulder, Colo.: Westview.

Anheier, Helmut K., and Lester M. Salamon. (1998) *The Non-profit Sector in the Developing World.* Manchester: Manchester University Press.

Anheier, Helmut K., and Stefan Toepler, eds. (1999) *Private Funds, Public Purpose: Philanthropic Foundations in International Perspective.* Dordrecht: Kluwer Academic.

Anti-Slavery Society. (1983) *The Philippines: Authoritarian Government, Multinationals and Ancestral Lands.* London: Anti-Slavery Society.

Arquiza, Yasmin D. (1996) "Palawan's Environmental Movement." In *Palawan at the Crossroads,* ed. James F. Eder and Janet O. Fernandez, 136–45. Manila: Ateneo de Manila University Press.

Badhwar, Neera Kapur. (1993) "Altruism versus Self-Interest: Sometimes a False Dichotomy." In *Altruism,* ed. Ellen Frankel Paul, Fred D. Miller Jr., and Jeffrey Paul, 90–117. Cambridge: Cambridge University Press.

Bailey, F. G., ed. (1971a) *Gifts and Poison.* Oxford: Basil Blackwell.

——. (1971b) "Gifts and Poison." In *Gifts and Poison,* ed. F. G. Bailey, 1–25. Oxford: Basil Blackwell.

Bankoff, Greg. (2003) *Cultures of Disaster: Society and Natural Hazard in the Philippines.* London: Routledge-Curzon.

Barrera, Esmi. (1990) "The Big Bird of Environmental Protection." *Philippine Star* (*Starweek*), 7 January.

Bar-Tal, D., R. Sharabany, and A. Raviv. (1982) "Cognitive Basis of the Development of Altruistic Behavior." In *Cooperation and Helping Behavior,* ed. V. J. Derlega and J. Grzelak, 377–96. London: Academic Press.

Batson, C. Daniel. (1991) *The Altruism Question*. Hillsdale: Lawrence Erlbaum.

Bauman, Z. (1993) *Postmodern Ethics*. Oxford: Blackwell.

Becker, Gary S. (1981) *A Treatise on the Family*. Cambridge: Harvard University Press.

Becker, Howard S. (1982) *Art Worlds*. Berkeley: University of California Press.

Beja, Edsel L. (1999) "Environmental Non-governmental Organizations in Policy-making: The Case against Lifting the Export Ban on Lumber in the Philippines." M.Phil. thesis, Department of Geography, University of Cambridge.

Belen, Angelo Ruel. (1996) Letter to the Hon. Fidel Ramos, 22 July.

——. (1998) "DENR Awards 'Ancestral land and Waters' to the Tagbanwa of Palawan." *Fieldnotes* n.v.: 13–15.

Benhabib, Seyla. (1992) *Situating the Self*. New York: Routledge.

Bennett, Jane. (2002) "The Moraline Drift." In *The Politics of Moralizing,* ed. Jane Bennett and Michael J. Shapiro, 11–26. New York: Routledge.

Bennett, Jon, and Sara Gibbs. (1996) *NGO Funding Strategies*. Oxford: INTRAC.

Blum, Lawrence A. (1980) *Friendship, Altruism and Morality*. London: Routledge and Kegan Paul.

Boudreau, Vincent G. (1996) "Of Motorcades and Masses: Mobilization and Innovation in Philippine Protest." In *The Revolution Falters,* ed. Patricio N. Abinales, 60–82. Ithaca: Southeast Asia Program, Cornell University.

——. (2001) *Grassroots and Cadre in the Protest Movement*. Quezon City: Ateneo de Manila University Press.

Bourdieu, Pierre. (1984) *Distinction*. Cambridge: Harvard University Press.

——. (1986) "The Forms of Capital." In *Handbook of Theory and Research for the Sociology of Education*, ed. John Richardson, 241–58. Westport: Greenwood Press.

Bourdieu, Pierre, and Loic J. D. Wacquant. (1992) *An Invitation to Reflexive Sociology*. Chicago: University of Chicago Press.

Boyce, James K. (1993) *The Political Economy of Growth and Impoverishment in the Marcos Era*. Quezon City: Ateneo de Manila University Press.

Braganza, Gilbert C. (1996) "Philippine Community-based Forest Management: Options for Sustainable Development." In *Environmental Change in South-East Asia*, ed. Michael J. G. Parnwell and Raymond L. Bryant, 311–29. London: Routledge.

Braun, Bruce, and Noel Castree, eds. (1998) *Remaking Reality: Nature at the Millennium*. London: Routledge.

Brett, E. A. (1993) "Voluntary Agencies as Development Organizations." *Development and Change* 24: 269–304.

Bridge, Gary. (2000) "Rationality, Ethics and Space." *Environment and Planning D: Society and Space* 18: 519–35.

——. (2001) "Bourdieu, Rational Action and the Time-Space Strategy of Gentrification." *Transactions of the Institute of British Geographers* 26: 205–16.

Broad, Robin, with John Cavanagh. (1993) *Plundering Paradise: The Struggle for the Environment in the Philippines*. Berkeley: University of California Press.

Brosius, J. Peter. (1999) "Anthropological Engagements with Environmentalism." *Current Anthropology* 40: 277–307.

Bryant, Raymond L. (2000) "Politicized Moral Geographies." *Political Geography* 19: 673–705.

——. (2001) "Explaining State-Environmental NGO Relations in the Philippines and Indonesia." *Singapore Journal of Tropical Geography* 22: 15–37.

——. (2002a) "Non-governmental Organizations and Governmentality: 'Consuming' Biodiversity and Indigenous People in the Philippines." *Political Studies* 50: 268–92.

——. (2002b) "False Prophets? Mutant NGOs and Philippine Environmentalism." *Society and Natural Resources* 15: 629–39.

Bryant, Raymond L., and Sinead Bailey. (1997) *Third World Political Ecology.* London: Routledge.

Bryant, Raymond L., and Michael Goodman. (2004) "Consuming Narratives: The Political Ecology of 'Alternative' Consumption." *Transactions of the Institute of British Geographers* 29: 344–66.

Bryant, Raymond L., and Lucy Jarosz, eds. (2004) "Ethics in Political Ecology." Special issue of *Political Geography* 23(7).

Cabague, Melanie. (1991) "Ecology and Survival in Caramoan." *Haribon Update* 6, no. 2 (March–April): 10.

Cala, Cesar. (1994) "Introduction." In *Studies on Coalition Experiences in the Philippines,* ed. Cesar Cala and Jose Grageda, 1–3. Manila: Bookmark.

Caucus of Development NGO Networks (CODE-NGO). (1995a) *Breakthroughs in Philippine Development*. Manila: CODE-NGO.

——. (1995b) *Scaling Up the Impact of Development NGOs in the Philippines*. Manila: CODE-NGO.

——. (1997) "NGO-managed Funding Mechanisms: A Formula for Genuine Partnership between NGOs and Donors?" *Philippine NGO Memo* 1 (January–March): 16–17.

Caucus of Development NGO Networks (CODE-NGO), Helvetas, and the Swiss Coalition of Development Organizations. (1995) *Building the Foundations of a Sustainable Society*. Manila: CODE-NGO.

Chaloupka, William. (2002) "The Tragedy of the Ethical Commons."

In *The Politics of Moralizing,* ed. Jane Bennett and Michael J. Shapiro, 113–40. New York: Routledge.

Chong, Dennis. (2000) *Rational Lives*. Chicago: University of Chicago Press.

Clad, James, and Marites Danguilan Vitug. (1988) "Philippines: The Plunder of Palawan." *Far Eastern Economic Review,* 24 November.

Clark, John. (1991) *Democratizing Development: The Role of Voluntary Organizations*. London: Earthscan.

Clarke, Gerard. (1998) *The Politics of NGOs in South-East Asia: Participation and Protest in the Philippines.* London: Routledge.

Colloredo-Mansfield, Rudi. (2002) "An Ethnography of Liberalism." *Current Anthropology* 43: 113–37.

Congress of the Philippines (CP). (1997) *The Indigenous Cultural Communities/Indigenous Peoples' Rights Act of 1997*. Manila: CP.

Conklin, Harold C. (1957) *Hanunóo Agriculture: A Report on an Integral System of Shifting Cultivation in the Philippines*. Rome: Food and Agriculture Organization.

Connell, Dan. (1991) "Palawan Endangered." *National Mid-Week,* 29 May.

Connell, J. (1999) "Beyond Manila: Walls, Malls, and Private Spaces." *Environment and Planning A* 31: 417–39.

Connelly, James, and Graham Smith. (2003) *Politics and the Environment: From Theory to Practice*. Second edition. London: Routledge.

Conservation International (CI). (1996) *Annual Report 1995*. Washington: CI.

——. (1997) *Philippines Program*. Manila: CI.

Constantino-David, Karina. (1992) "The Philippine Experience in Scaling-up." In *Making a Difference,* ed. Michael Edwards and David Hulme, 137–47. London: Earthscan.

——. (1997) "Intra-Civil Society Relations: A Synoptic Paper." Manila: University of the Philippines Unpublished MS.

——. (1998) "From the Present Looking Back." In *Organizing for Democracy*, ed. G. Sidney Silliman and Lela Garner Noble, 26–48. Honolulu: University of Hawaii Press.

Corbridge, Stuart. (1993) "Marxisms, Modernities and Moralities: Development Praxis and the Claims of Distant Strangers." *Environment and Planning D: Society and Space* 11: 449–72.

Cox, Kevin R. (2003) "Political Geography and the Territorial." *Political Geography* 22: 607–10.

Dalton, Russell J. (1994) *The Green Rainbow*. New Haven: Yale University Press.

Dauvergne, Peter. (1997) *Shadows in the Forest: Japan and the Politics of Timber in Southeast Asia*. Cambridge, Mass.: MIT Press.

Della Porta, Donatella, and Mario Diani. (1999) *Social Movements*. Oxford: Blackwell.

DeLuca, Kevin Michael. (1999) *Image Politics*. New York: Guilford Press.

Department of Environment and Natural Resources (DENR). (1992) "Department Administrative Order (DAO) 52 of 1992." Manila: DENR.

——. (1994) "Guidelines for the Watchlisting and Blacklisting of DENR's Accredited NGOs and Contractors." Manila: DENR.

——. (1995) *Memorandum between the DENR and the PAFID*. Manila: DENR.

——. (1996a) "Guidelines on the Management of Certified Ancestral Domain Claims." Manila: DENR.

de Quiros, Conrado. (1996a) "Defending Paradise." *Philippine Daily Inquirer*, 12 August.

——. (1996b) "Sacred Cows." *Philippine Daily Inquirer*, 13 August.

de Rosas-Ignacio, Anna. (1997) "The Fundamentals of Sustainability." *Philippine NGO Memo* 3–4 (July–December): 3.

De Vera, Dave. (1994) "Low Income Upland Communities Project."

In *NGO-GO Relationship in Upland Development,* ed., Upland NGO Assistance Committee (UNAC), 39–50. Manila: UNAC.

Dicken, Peter. (2003) *Global Shift: Reshaping the Global Economic Map in the 21st Century*. Fourth Edition. London: Sage.

Dickson, Lisa, and Alistair McCulloch. (1996) "Shell, the Brent Spar and Greenpeace: A Doomed Tryst?" *Environmental Politics* 5: 122–29.

Dobson, Andrew. (2000) *Green Political Thought*. Third Edition. New York: Routledge.

Dolor, Beting Laygo, Romulo Luib, Elisha Garcia, and Io Aceremo. (1994a) "Code of Ethics Is in Order." *Business World,* 24 August.

——. (1994b) "Who Monitors Them?" *Business World,* 19 August.

Doyle, James. (1998) "Power and Contentment." *Politics* 18: 49–56.

Doyle, Timothy, and Doug McEachern. (2001) *Environment and Politics.* Second Edition. New York: Routledge.

Doyo, Ceres. (1997) "Bolinao Revisited." *Philippine Daily Inquirer,* 25 May.

Dreyfus, Hubert, and Paul Rabinow. (1999) "Can There Be a Science of Existential Structure and Social Meaning?" In *Bourdieu: A Critical Reader,* ed. Richard Shusterman, 84–93. Oxford: Blackwell.

Eccleston, Bernard, and David Potter. (1996) "Environmental NGOs and Different Political Contexts in South-East Asia." In *Environmental Change in South-East Asia,* ed. Michael Parnwell and Raymond Bryant, 49–66. London: Routledge.

Ecologist. (1993) *Whose Common Future?* London: Earthscan.

Edelman, Murray. (1974) "The Political Language of the 'Helping Professions.'" *Politics and Society* 4: 295–310.

Eder, Klaus. (1993) *The New Politics of Class*. London: Sage.

Edwards, Michael, and David Hulme, eds. (1992) *Making a Difference.* London: Earthscan.

Ekins, Paul. (1992) *A New World Order.* London: Routledge.

Eldridge, Philip. (1995) *Non-government Organizations and Democratic Participation in Indonesia*. Kuala Lumpur: Oxford University Press.

Emmons, Karen. (1997) "Seahorse Samaritan." *Far Eastern Economic Review*, 11 September, p. 74.

Environmental Management Bureau. (1996) *Philippine Environmental Quality Report 1990–1995*. Manila: DENR.

Environmental Science for Social Change (ESSC). (1999) *Mining Revisited: Can an Understanding of Perspectives Help?* Manila: ESSC.

Espeland, Wendy Nelson. (1998) *The Struggle for Water*. Chicago: University of Chicago Press.

Eyerman, Ron, and Andrew Jamison. (1991) *Social Movements: A Cognitive Approach*. Cambridge: Polity.

Fabros, Wilfredo. (1988) *The Church and Its Social Involvement in the Philippines, 1930–1972*. Manila: Ateneo de Manila University Press.

Farrington, John, and Anthony Bebbington. (1993) *Reluctant Partners? Non-governmental Organizations, the State and Sustainable Agricultural Development*. London: Routledge.

Ferree, Myra Marx. (1992) "The Political Context of Rationality: Rational Choice Theory and Resource Mobilization." In *Frontiers in Social Movement Theory*, ed. Aldon D. Morris and Carol McClurg Mueller, 29–52. New Haven: Yale University Press.

Fine, Ben. (2001) *Social Capital versus Social Theory*. London: Routledge.

Fine, Gary Alan. (2001) *Difficult Reputations*. Chicago: University of Chicago Press.

Fisher, Julie. (1993) *The Road from Rio: Sustainable Development and the Non-governmental Movement in the Third World*. Westport, Conn.: Praeger.

——. (1998) *Nongovernments: NGOs and the Political Development of the Third World*. West Hartford: Kumarian Press.

Fisher, William F. (1997) "Doing Good? The Politics and Antipolitics of NGO Practices." *Annual Review of Anthropology* 26: 439–64.

Ford Foundation. (1997) *Ford Foundation in the Philippines*. Manila: Ford Foundation.

Forsyth, Tim. (2003) *Critical Political Ecology*. New York: Routledge.

Foundation for the Philippine Environment (FPE). (1995) *1994 Annual Report*. Manila: FPE.

——. (1996) *1995 Annual Report*. Manila: FPE.

Foweraker, Joe. (1995) *Theorizing Social Movements*. London: Pluto Press.

Fowler, Alan. (1997) *Striking a Balance*. London: Earthscan.

——. (2000) *The Virtuous Spiral*. London: Earthscan.

French, Marilyn. (1985) *Beyond Power*. New York: Ballantine.

Fultz, J., and R. B. Cialdini. (1991) "Situational and Personality Determinants of the Quantity and Quality of Helping." In *Cooperation and Prosocial Behaviour,* ed. Robert A. Hinde and Jo Groebel, 135–46. Cambridge: Cambridge University Press.

Gabor, Mina. (1996) Letter to Ruel Belen, 7 August.

Ganapin, Delfin. (1989) "Strategic Assessment of NGOs in Community Forestry and Environment." In *A Strategic Assessment of Nongovernmental Organizations in the Philippines,* ed. Antonio B. Quizon and Rhoda U. Reyes, 85–101. Manila: Asian NGOs Coalition for Agrarian Reform and Rural Development.

Gasgonia, Donna. (1997) "Ancestral Islands and Ancestral Waters." Manila: mimeo.

Gauld, Richard. (2000) "Maintaining Centralized Control in Community-based Forestry." *Development and Change* 31: 229–54.

Giddens, Anthony. (1984) *The Constitution of Society*. Berkeley: University of California Press.

Gillroy, John Martin, and Joe Bowersox, eds. (2002) *The Moral Austerity of Environmental Decision Making*. Durham, N.C.: Duke University Press.

Glassman, James F. (2003) "Structural Power, Agency and National Liberation: The Case of East Timor." *Transactions of the Institute of British Geographers* 28: 264–80.

Go, Julian, and Anne L. Foster, eds. (2003) *The American Colonial State in the Philippines: Global Perspectives*. Durham, N.C.: Duke University Press.

Goodin, Robert E. (1992) *Motivating Political Morality*. Oxford: Blackwell.

Goodman, David, and Mike Goodman. (2001) "Sustaining Foods: Organic Consumption and the Socio-ecological Imaginary." In *Exploring Sustainable Consumption: Environmental Policy and the Social Sciences*, ed. M. Cohen and J. Murphy, 97–119. Oxford: Elsevier.

Goodno, James B. (1989) *The Philippines: Land of Broken Promises*. London: Zed.

Goodwin, Jeff, James M. Jasper, and Francesca Polletta, eds. (2001) *Passionate Politics: Emotions and Social Movements*. Chicago: University of Chicago Press.

Goverde, Henri, Philip G. Cerny, Mark Haugaard, and Howard Lentner. (2000) "Power in Contemporary Politics." In *Power in Contemporary Politics*, ed. Henri Goverde, Philip G. Cerny, Mark Haugaard, and Howard Lentner, 1–33. London: Sage.

Grant, Ruth W. (1997) *Hypocrisy and Integrity: Machiavelli, Rousseau, and the Ethics of Politics*. Chicago: University of Chicago Press.

Gray, M. L. (1999) "Creating Civil Society? The Emergence of NGOs in Vietnam." *Development and Change* 30: 693–713.

Green Forum. (1993) "The Philippine Anti-Logging Movement." Manila: mimeo.

Gregorio-Medel, A. (1993) "Development Work Is Middle Class Oriented as Much as It Is Poverty Oriented." *Philippine Politics and Society* (January): 58–93.

Gudeman, Stephen. (2001) *The Anthropology of Economy: Community, Market, and Culture*. Oxford: Blackwell.

Gupta, Akhil, and James Ferguson. (1997) "Culture, Power, Place: Ethnography at the End of an Era." In *Culture, Power, Place: Explorations in Critical Anthropology*, ed. Akhil Gupta and James Ferguson, 1–29. Durham, N.C.: Duke University Press.

Guralnik, David B., ed. (1986) *Webster's New World Dictionary of the American Language*. Second College Edition. New York: Simon and Schuster.

Gutierrez, Eric. (1994) *The Ties that Bind*. Manila: Philippine Center for Investigative Journalism.

Habermas, Jürgen. (1984) *The Theory of Communicative Action*. Volume 1. Boston: Beacon.

Hakim, Catherine. (1987) *Research Design*. London: Unwin Hyman.

Hamel, Jacques. (1993) *Case Study Methods*. London: Sage.

Hannigan, John A. (1995) *Environmental Sociology*. London: Routledge.

Haribon Foundation (HF). (1987) *Update* (November).

——. (1988) *Update* (September–November).

——. (1989a) *Update* (January–February).

——. (1989b) "Communal Forest Lease (CFL) Application for Palaweños." *Haribon Update* (September–October), 4.

——. (1989c) *Haribon Update* (September–October).

——. (1989d) "Gorillas in the Mist." Manila: Mimeo of newspaper advertisement.

——. (1990a) *Haribon Update* (January–February).

——. (1990b) *Haribon Update* (May–June).

——. (1990c) *Haribon Update* (September–October).

——. (1990d) *Haribon Update* (November–December).

——. (1991a) *Haribon Update* (March–April).

——. (1991b) *Haribon Update* (May–June).

——. (1991c) *Haribon Update* (September–October).

——. (1992) *Haribon Update* (May–June).

——. (1995) *Haribon Update* (July–September).

——. (1989–1996) *Minutes of the Board of Trustees*. Manila: HF.

——. (1996a) *Haribon Quarterly* (April–September).

——. (1996b) "Membership Form." Manila: HF.

——. (1997a) *1996 Year End Report*. Manila: HF.

——. (1997b) "Description of Ongoing Projects." Manila: HF.

Haribon Society. (1982) *The Hariboner* (November–December).

——. (1984) *The Hariboner* (January).

Hart, Stephen. (2001) *Cultural Dilemmas of Progressive Politics: Styles of Engagement among Grassroots Activists*. Chicago: University of Chicago Press.

Harvey, David. (1996) *Justice, Nature and the Geography of Difference*. Oxford: Blackwell.

——. (1998) "The Body as Accumulation Strategy." *Environment and Planning D: Society and Space* 16: 401–21.

Hay, Colin. (1997) "Divided by a Common Language: Political Theory and the Concept of Power." *Politics* 17: 45–52.

Heyzer, Noeleen, James V. Riker, and Antonio B. Quizon, eds. (1995) *Government-NGO Relations in Asia*. London: Macmillan.

Hilhorst, Dorothea. (1997) "Discourse Formation in Social Movements." In *Images and Realities of Rural Life*, ed. H. de Haan and Norman Long, 121–49. Assen: Van Gorcum.

——. (2003) The Real World of NGOs: Discourses, Diversity and Development. London: Zed Press.

Hjelmar, Ulf. (1996) *The Political Practice of Environmental Organizations*. Aldershot: Avebury.

Honasan, Alya B. (1996) "Places in the Sun." *Sunday Inquirer Magazine*, 10 November.

Howell, Signe, ed. (1997) *The Ethnography of Moralities*. London: Routledge.

Hughes, A. (2001) "Global Commodity Networks, Ethical Trade and Governmentality: Organizing Business Responsibility in the Kenyan Cut Flower Industry." *Transactions of the Institute of British Geographers* 26: 390–406.

Hugman, Richard. (1991) *Power in the Caring Professions*. London: Macmillan.

Hulme, David, and Michael Edwards, eds. (1997a) *NGOs, States and Donors: Too Close for Comfort?* London: Macmillan.

Hulme, David, and Michael Edwards. (1997b) "Conclusion." In *NGOs, States and Donors: Too Close for Comfort*, ed. David Hulme and Michael Edwards, 275–84. London: Macmillan.

Ibarra, Peter R., and John I. Kitsuse. (1993) "Vernacular Constituents of Moral Discourse." In *Reconsidering Social Constructionism*, ed. James A. Holstein and Gale Miller, 25–58. New York: Aldine de Gruyter.

Jaggar, Allison. (1983) *Feminist Politics and Human Nature*. Totowa: Rowman and Allanheld.

Jakobeit, Cord. (1996) "Nonstate Actors Leading the Way: Debt-for-Nature Swaps." In *Institutions for Environmental Aid*, ed. Robert Keohane and Marc Levy, 127–66. Cambridge, Mass.: MIT Press.

Jaravelo, Fredo, and Willy Tolentino. (1989) "Travel Report: Coron Island, February 25 to March 2, 1989." Manila: PAFID.

Jasper, James. (1997) *The Art of Moral Protest*. Chicago: University of Chicago Press.

Jimenez, Pol. (1993) "Funds Infused to NGOs Have Little Impact." *Philippine Star*, 24 October.

Johnson, Peter. (1993) *Frames of Deceit*. Cambridge: Cambridge University Press.

Johnston, Hank, and Bert Klandermans, eds. (1995) *Social Movements and Culture*. London: UCL Press.

Johnston, Ron. (2001) "Out of the 'Moribund Backwater': Territory and Territoriality in Political Geography." *Political Geography* 20: 677–94.

Johnston, R. J., Derek Gregory, and David M. Smith, eds. (1994) *The Dictionary of Human Geography*. Third Edition. Oxford: Blackwell.

Johnston, R. J., Peter J. Taylor, and Michael J. Watts, eds. (2002) *Geographies of Global Change: Remapping the World*. Oxford: Blackwell.

Jordan, Bill. (1989) *The Common Good*. Oxford: Basil Blackwell.

Jordan, Grant, and William Maloney. (1997) *The Protest Business*. Manchester: Manchester University Press.

Kalaw, Maximo T. (1989) "Democratizing Filipino Resources." Manila: mimeo.

——. (1990) "Sustainable Development and USAID: The Philippine Experience." *Haribon Update* (September–October), 8.

——. (1997) *Exploring Soul and Society*. Manila: Anvil.

Kane, John. (2001) *The Politics of Moral Capital*. Cambridge: Cambridge University Press.

Kaplan, C. (1995) "'A World without Borders': The Body Shop's Trans/national Geographies." *Social Text* 13: 45–66.

Karacs, Imre. (1999) "Greenpeace Boss Told: On Your Bike." *The Independent on Sunday*, 7 February.

Keck, Margaret E., and Kathryn Sikkink. (1998) *Activists beyond Borders: Advocacy Networks in International Politics*. Ithaca: Cornell University Press.

Kellow, Aynsley. (2000) "Norms, Interests and Environment NGOs: The Limits of Cosmopolitanism." *Environmental Politics* 9 (Autumn): 1–22.

Kelly, Philip F. (2002) "Spaces of Labor Control: Comparative Perspectives from Southeast Asia." *Transactions of the Institute of British Geographers* 27: 395–411.

Kitschelt, Herbert. (1986) "Political Opportunity Structures and Political Protest." *British Journal of Political Science* 16: 57–85.

Klein, Daniel B., ed. (1997) *Reputation: Studies in the Voluntary Elicitation of Good Conduct*. Ann Arbor: University of Michigan Press.

Klein, Naomi. (2000) *No Logo*. London: Flamingo.

Korsgaard, Christine M. (1993) "The Reasons We Can Share." In *Altruism*, ed. Ellen Frankel Paul, Fred D. Miller Jr., and Jeffrey Paul, 24–51. Cambridge: Cambridge University Press.

Korten, David. (1990) *Getting to the 21st Century.* West Hartford: Kumarian Press.

Korten, Francis. (1994) "Questioning the Call for Environmental Loans: A Critical Examination of Forestry Lending in the Philippines." *World Development* 22: 971–81.

Kummer, David M. (1992) *Deforestation in the Postwar Philippines.* Manila: Ateneo de Manila University Press.

Kurtz, Hilda E. (2003) "Scale Frames and Counter-Scale Frames: Constructing the Problem of Environmental Injustice." *Political Geography* 22: 887–916.

Lahusen, Christian. (1996) *The Rhetoric of Moral Protest.* New York: Walter de Gruyter.

Lang, Gladys Engel, and Kurt Lang. (1990) *Etched in Memory: The Building and Survival of Artistic Reputation*. Chapel Hill: University of North Carolina Press.

Lara, Ben. (1974) "Conduct Studies on Environment." *Bulletin Today,* 24 March.

Lara, Maria Pia. (1998) *Moral Textures*. Cambridge: Polity.

Lawrence, Karen. (2002) "Negotiated Biodiversity Conservation for Local Social Change: A Case Study of Northern Palawan, Philippines." Ph.D. thesis. Department of Geography, King's College London.

Lean, Geoffrey. (1997) "BP Occupies a Greenpeace Platform." *Independent on Sunday,* 12 October.

Lee, Jeng-di. (2004) "Managing coastal resources in the Philippines." Ph.D. thesis. Department of Geography, King's College London.

Legazpi, Eileen. (1994) "Environmental Coalitions." In *Studies on Coalition Experiences in the Philippine,* ed. Cesar Cala and Jose Grageda, 121–53. Manila: Bookmark.

Leonen, Marvic. (2000) "NGO Influence on Environmental Policy." In *Forest Policy and Politics in the Philippines,* ed. Peter Utting, 67–83. Quezon City: Ateneo de Manila University Press.

Libre, Shirley (PAFID Executive Director). (1993). Letter to DENR Secretary Angel Alcala, November 25.

Lin, Nan. (2001) *Social Capital: A Theory of Social Structure and Action*. Cambridge: Cambridge University Press.

Lopa, Consuela Katrina A., and Karel S. San Juan. (1996) "NGO Relations with Donor Agencies." In *Trends and Traditions, Challenges and Choices,* ed. Alan Alegre, 128–36. Manila: Ateneo Center for Social Policy and Public Affairs.

Lopez, Amor. (1991) "Haribon, Others Infiltrated by Reds?" *Manila Bulletin,* 18 March.

Low Income Upland Communities Project. (1993) *Minutes of the NGO Development Workshop*. N.p.: PAFID mimeo.

Low, Nicholas, and Brendan Gleeson. (1998) *Justice, Society and Nature: An Exploration of Political Ecology*. New York: Routledge.

Lozada, Becky. (1997) "Blood Compact to Stop Bid for Cement Plant." *Philippine Daily Inquirer,* 16 May.

Lukes, Steven. (1974) *Power: A Radical View*. London: Methuen.

——. (1991) *Moral Conflict and Politics*. Oxford: Clarendon.

Marco, Jesusa. (1994) *The Low Income Upland Communities Project*. Manila: De La Salle University.

Martin, Deborah G. (2003) " 'Place-Framing' as Place-Making: Constituting a Neighborhood for Organizing and Activism." *Annals of the Association of American Geographers* 93: 730–50.

McAdam, Doug, John D. McCarthy, and Mayer N. Zald, eds. (1996) *Comparative Perspectives on Social Movements.* Cambridge: Cambridge University Press.

McAfee, Kathleen. (1999) "Selling Nature to Save It? Biodiversity and Green Developmentalism." *Environment and Planning D: Society and Space* 17: 133–54.

McCarthy, John D., and Mayer N. Zald. (1973) *The Dynamics of Social Movements: Resource Mobilization, Social Control and Tactics.* Cambridge, Mass.: Winthrop Publishers.

——. (1987) "The Trend of Social Movements in America." In *Social Movements in an Organizational Society,* ed. Mayer Zald and John McCarthy, 337–92. New Brunswick: Transaction.

Melucci, Alberto. (1989) *Nomads of the Present.* Philadelphia: Temple University Press.

——. (1996) *Challenging Codes.* Cambridge: Cambridge University Press.

Mercer, Claire. (2002) "NGOs, Civil Society and Democratization." *Progress in Development Studies* 2: 5–22.

——. (2003) "Performing Partnership: Civil Society and the Illusions of Good Governance in Tanzania." *Political Geography* 22: 741–63.

Mercer, Jonathan. (1996) *Reputation and International Politics.* Ithaca: Cornell University Press.

Meyer, Carrie. (1995) "Opportunism and NGOs: Entrepreneurship and Green North-South Transfers." *World Development* 23: 1277–89.

——. (1999) *The Economics and Politics of NGOs in Latin America.* Westport, Conn.: Praeger.

Miller, Byron. (1992) "Collective Action and Rational Choice: Place, Community and the Limits to Individual Self-interest." *Economic Geography* 68: 22–42.

Moorhead, Caroline. (1998) *Dunant's Dream: War, Switzerland and the History of the Red Cross.* London: Harper Collins.

Nagel, Thomas. (1970) *The Possibility of Altruism*. Princeton: Princeton University Press.

Najam, Adil. (1996) "NGO Accountability: A Conceptual Framework." *Development Policy Review* 14: 339–53.

National Commission on Indigenous Peoples. (1998) *Rules and Regulations Implementing the "Indigenous Peoples" Rights Act of 1997*. Manila: Office of the President.

National Integrated Protected Areas Program (NIPAP). (1997) *Annual Report for 1996*. Manila: NIPAP.

——. (1999) "NIPAP Gains Tagbanua Support in Coron Island." *U.L.A.T. NIPAP* (NIPAP Newsletter) 2, no. 3 (June): 1–2.

Navarro, Karayan. (1995) "New Hopes for an Ancient Mountain." *Haribon Quarterly* (October–December), 12–14.

Netherlands Embassy. (1997) "Local Environmental Fund." Manila: Netherlands Embassy.

Netherlands Government. (1997) "Current Environmental Involvements in the Philippines." Manila: Netherlands Government.

Nettleton, Geoff. (1996a) "Constitution Undermined by Mining Code." *Higher Values: The Minewatch Bulletin* 9: 18–19.

——. (1996b) "Mining Gang Flex Their Muscles in the Philippines." *Higher Values: The Minewatch Bulletin* 10: 18–19.

NGOs for Integrated Protected Areas (NIPA). (1996) "A Case Study on NGO-GO-WB Dynamics in a GEF Funded Philippine Integrated Areas Project." Manila: NIPA.

——. (1997) *Information Kit*. Manila: NIPA.

Noering, Carolyn. (1982) "The Batanes Adventure." *The Hariboner* 1 (2): 2.

Nussbaum, Martha C. (1986) *The Fragility of Goodness*. Cambridge: Cambridge University Press.

Nuyda, Doris Gaskell. (1997a) "A Look at the 'New' Haribon." *Philippine Daily Inquirer*, 28 February.

——. (1997b) "Haribon Scores a Victory for the Environment." *Philippine Daily Inquirer,* 7 March.

O'Connor, James. (1998) *Natural Causes.* New York: Guilford Press.

Ong, Aihwa. (1999) *Flexible Citizenship: The Cultural Logics of Transnationality.* Durham, N.C.: Duke University Press.

Pajaro, Marivic G. (1992) "Alternative to Sodium Cyanide Use in Aquarium Fish Collection: A Community-based Approach." *Enviroscope: A Bulletin of the Haribon Foundation* 7, no. 7: 1–12.

Parry, J., and M. Bloch, eds. (1989) *Money and the Morality of Exchange.* Cambridge: Cambridge University Press.

Paul, Ellen Frankel, Fred D. Miller Jr., and Jeffrey Paul, eds. (1997) *Self-Interest: Part I.* Cambridge: Cambridge University Press.

Peet, Richard, and Michael Watts, eds. (2004) *Liberation Ecologies.* Second Edition. London: Routledge.

Philipino Express. (1973) "Editorial: Caring for Nature." 2 July, p. 6.

Philippine Association for Intercultural Development (PAFID). (1977–1978) *Minutes of the Board of Trustees.* Manila: PAFID.

——. (1984) "Draft of PAFID Quarterly Report for January 1–March 31, 1984." Manila: PAFID mimeo.

——. (1987) "Field Trip Report: Langawa, Ifugao (April 15–28, 1987)." Manila: PAFID.

——. (1991) "Statement of Funds and Disbursements as of December 31, 1990." Manila: PAFID mimeo.

——. (1992–1996) *Minutes of the Board of Trustees.* Manila: PAFID.

——. (1993a) *Communal Title: A Valid Option for Land Tenure for Tribal Filipinos?* Manila: PAFID.

——. (1993b) "25 Years of Partnership with Indigenous Filipinos." *Fieldnotes* 8, no. 1 (June–September): 1–12.

——. (1993c) *Proceedings of the PAFID Year-end Assessment, 15–17 December 1992.* Manila: PAFID.

——. (1994a) "Audited Financial Statements December 31, 1993." Manila: PAFID.

——. (1994b) *Survey and Delineation of Ancestral Lands and Domains (BSP First Interim Report)*. Manila: PAFID.

——. (1994c) *Annual Report (1993–1994)*. Manila: PAFID.

——. (n.d.) *Progress Report on Mangyan Self-Sufficiency Project*. Manila: PAFID.

——. (1995a) *The PAFID Land Tenure and Community Development Program for Tribal Filipinos*. Manila: PAFID.

——. (1995b) *Progress Report to Misereor for September 1994–February 1995*. Manila: PAFID.

——. (1995c) *Progress Report to Misereor, March–August 1995*. Manila: PAFID.

——. (n.d.) *Human Resource Management Manual*. Manila: PAFID.

——. (1998) *Mapping the Ancestral Lands and Waters of the Calamian Tagbanwa*. Manila: PAFID.

Philippine Council for Sustainable Development (PCSD). (1996) *Philippine Agenda 21: A National Agenda for Sustainable Development for the 21st Century*. Manila: PCSD.

Philippine Daily Inquirer. (1991a) "PNP Rapped for Arrest of Haribon Staff." 22 February.

——. (1991b) "Mitra Assails Haribon for 'Smear' Drive." 3 March.

——. (1996) "Bolinao community wins." 9 August.

Philippine Partnership for the Development of Human Resources in Rural Areas (PhilDHRRA). (1996) *Commitment to the Future*. Manila: PhilDHRRA.

——. (1997) *1995 Annual Report and 1996 Social Accounts*. Manila: PhilDHRRA.

Philippine Rural Reconstruction Movement (PRRM). (1996) *Building Community and Habitat*. Manila: PRRM.

Plan International. (1996a) *Worldwide Annual Report 1996*. London: Westdale.

——. (1996b) *Annual Report: Philippines (July 1995–June 1996)*. Manila: Plan International.

——. (1996c) *An Introduction to Plan International.* London: Plan International.

Plant, Raymond. (1992) "Enterprise in Its Place: The Moral Limits of Markets." In *The Values of the Enterprise Culture,* ed. Paul Heelas and Paul Morris, 85–99. London: Routledge.

Polletta, Francesca. (2002) *Freedom Is an Endless Meeting: Democracy in American Social Movements.* Chicago: University of Chicago Press.

Portes, Alejandro. (1998) "Social Capital: Its Origins and Applications in Modern Sociology." *Annual Review of Sociology* 24: 1–24.

Prakash, A. (2000) *Greening the Firm: The Politics of Corporate Environmentalism.* Cambridge: Cambridge University Press.

Princen, Thomas, and Matthias Finger. (1994) *Environmental NGOs in World Politics.* London: Routledge.

Putnam, Robert. (1993) *Making Democracy Work.* Princeton: Princeton University Press.

——. (2000) *Bowling Alone.* New York: Simon and Schuster.

Putzel, James. (1992) *A Captive Land.* Manila: Ateneo de Manila University Press.

Radin, Margaret Jane. (1996) *Contested Commodities.* Cambridge: Harvard University Press.

Rampton, Sheldon, and John Stauber. (2001) *Trust Us, We're Experts! How Industry Manipulates Science and Gambles with Your Future.* New York: Jeremy P. Tarcher.

Raquiza, Toinette. (n.d.) "Making a Stand for the Environment." *Rural Reconstruction Forum* 2, no. 3: 1–2.

Repetto, Robert, and Malcolm Gillis, eds. (1988) *Public Policies and the Misuse of Forest Resources.* Cambridge: Cambridge University Press.

Residents of Mapayao (Nueva Vizcaya). (1992) Letter to the PAFID Board of Directors, 10 June.

Richter, Judith. (2001) *Holding Corporations Accountable: Corporate Conduct, International Codes and Citizen Action*. London: Zed Press.

Riipinen, Pasi. (1995) *Making an Alternative Living in the Philippines*. Helsinki: Finnish Philippine Society.

Robinson, Mark. (1997) "Privatizing the Voluntary Sector: NGOs as Public Service Contractors?" In *NGOs, States and Donors: Too Close for Comfort?*, ed. David Hulme and Michael Edwards, 59–78. London: Macmillan.

Rocamora, Joel. (1994) *Breaking Through*. Manila: Anvil.

Romero, Andre. (1996) "Some Comments on Bruce Young's Outline of Proposed PAFID Mapping Section." Manila: PAFID.

Rood, Steven. (1998) "NGOs and Indigenous Peoples." In *Organizing for Democracy*, ed. G. Sidney Silliman and Lela Garner Noble, 138–56. Honolulu: University of Hawaii Press.

Roque, Anselmo. (1997) "Ikalahan Tribesmen Serve as Environmentalist Model." *Philippine Daily Inquirer*, 23 March and 24 March.

Ross, Michael. (1996) "Conditionality and Logging Reform in the Tropics." In *Institutions for Environmental Aid*, ed. Robert O. Keohane and Marc A. Levy, 177–97. Cambridge, Mass.: MIT Press.

——. (2001) *Timber Booms and Institutional Breakdown in Southeast Asia*. Cambridge: Cambridge University Press.

Routledge, Paul. (2003) "Convergence Space: Process Geographies of Grassroots Globalization Networks." *Transactions of the Institute of British Geographers* 28: 333–49.

Ruiz, Lester. (1990) "Sovereignty as Transformative Practice." In *Contending Sovereignties*, ed. R. B. J. Walker and Saul Mendlovitz, 79–96 . Boulder, Colo.: Lynne Rienner.

Sack, Robert D. (1986) *Human Territoriality*. Cambridge: Cambridge University Press.

——. (1997) *Homo Geographicus: A Framework for Action, Awareness, and Moral Concern*. Baltimore: Johns Hopkins University Press.

Salamon, Lester. (1994) “The Rise of the Nonprofit Sector.” *Foreign Affairs* 73, no. 4: 109–22.

Sayer, Andrew. (2000) “Critical and Uncritical Cultural Turns.” In *Cultural Turns/Geographical Turns,* ed. Ian Cook, David Crouch, Simon Naylor, and James R. Ryan, 166–81. London: Prentice Hall.

Schlosberg, David. (1999) *Environmental Justice and the New Pluralism.* Oxford: Oxford University Press.

Schlosser, Eric. (2001) *Fast Food Nation*. London: Allen Lane.

Schmidtz, David. (1993) “Reasons for Altruism.” In *Altruism,* ed. Ellen Frankel Paul, Fred D. Miller Jr., and Jeffrey Paul, 52–68. Cambridge: Cambridge University Press.

Schoenberger, Erica. (1998) “Discourse and Practice in Human Geography.” *Progress in Human Geography* 22 (March): 1–14.

Scott, Alan. (1990) *Ideology and the New Social Movements*. London: Unwin Hyman.

Scott, James C. (1985) *Weapons of the Weak*. New Haven: Yale University Press.

——. (1990) *Domination and the Arts of Resistance: Hidden Transcripts*. New Haven: Yale University Press.

Serrano, Isagani R. (1994) *Pay Now, Not Later*. Manila: Philippine Rural Reconstruction Movement.

Severino, Horacio. (1993) “Ormoc Revisited.” In *Saving the Earth,* Third Edition, ed. Eric Gamalinda and Sheila Coronel, 48–53. Manila: Philippine Center for Investigative Journalism.

——. (1995) “Fallen Angel.” *I: The Investigative Reporting Magazine* 1, no. 3 (July–September): 6–11.

Sidel, John T. (1999) *Capital, Coercion, and Crime*. Stanford: Stanford University Press.

Silliman, G. Sidney, and Lela Garner Noble, eds. (1998a) *Organizing for Democracy*. Honolulu: University of Hawaii Press.

——. (1998b) “Introduction.” In *Organizing for Democracy,* ed. G.

Sidney Silliman and Lela Garner Noble, 3–25. Honolulu: University of Hawaii Press.

Sinclair, Timothy J. (2000) "Reinventing Authority: Embedded Knowledge Networks and the New Global Finance." *Environment and Planning C* 18: 487–502.

Skillen, Tony. (1992) "Enterprise: Towards the Emancipation of a Concept." In *The Values of the Enterprise Culture,* ed. Paul Heelas and Paul Morris, 73–82. London: Routledge.

Slim, Hugo. (1997) "To the Rescue: Radicals or Poodles?" *The World Today* (London) August–September: 209–12.

Smillie, Ian. (1995) *The Alms Bazaar.* London: Intermediate Technology.

Smillie, Ian, and John Hailey. (2001) *Managing for Change: Leadership, Strategy and Management in Asian NGOs.* London: Earthscan.

Smith, Brian H. (1990) *More than Altruism: The Politics of Private Foreign Aid.* Princeton: Princeton University Press.

Smith, David M. (1998) "How Far Should We Care? On the Spatial Scope of Beneficence." *Progress in Human Geography* 22: 15–38.

——. (1999) "Geography and Ethics: How Far Should We Go?" *Progress in Human Geography* 23: 119–25.

——. (2000) *Moral Geographies: Ethics in a World of Difference.* Edinburgh: Edinburgh University Press.

Smith, Joe. (2000) "After the Brent Spar: Business, the Media and the New Environmental Politics." In *The Daily Globe,* ed. Joe Smith, 168–85. London: Earthscan.

Snow, David A., and Robert D. Benford. (1992) "Master Frames and Cycles of Protest." In *Frontiers in Social Movement Theory,* ed. Aldon D. Morris and Carol McClurg Mueller, 133–55. New Haven: Yale University Press.

Snow, David, and Doug McAdam. (2000) "Identity Work Processes in the Context of Social Movements." In *Self, Identity, and Social*

Movements, ed. Sheldon Stryker, Timothy J. Owens, and Robert W. White, 41–67. Minneapolis: University of Minnesota Press.

Snow, David A., E. Burke Rochford Jr., Steven K. Worden, and Robert D. Benford. (1986) "Frame Alignment Processes, Micromobilization, and Movement Participation." *American Sociological Review* 51: 464–81.

Sogge, David, ed. (1996a) *Compassion and Calculation.* London: Pluto Press.

——. (1996b) "Settings and Choices." In *Compassion and Calculation,* ed. David Sogge, 1–23. London: Pluto Press.

Stauber, John, and Sheldon Rampton. (1995) " 'Democracy' for Hire: Public Relations and Environmental Movements." *The Ecologist* 25: 173–80.

Storey, David. (2001) *Territory: The Claiming of Space.* London: Prentice Hall.

Sundberg, Juanita. (2003) "Conservation and Democratization: Constituting Citizenship in the Maya Biosphere Reserve, Guatemala." *Political Geography* 22: 715–40.

Swartz, Davis. (1997) *Culture and Power: The Sociology of Pierre Bourdieu.* Chicago: University of Chicago Press.

Swyngedouw, Erik. (1997) "Neither Global nor Local: Globalization and the Politics of Scale." In *Spaces of Globalization,* ed. Kevin Cox, 137–66. New York: Guilford.

Tagbanua Foundation Coron Island (TFCI). (1996) "Ancestral Domain Management Plan." N.p.: mimeo.

Tan, Jose Lorenzo. (1996) "Palawan Wildlife." *Mabuhay* (September): 36–41.

Tandon, Rajesh. (1995) " 'Board Games': Governance and Accountability in NGOs." In *Non-governmental Organizations: Performance and Accountability,* ed. Michael Edwards and David Hulme, 41–49. London: Earthscan.

Tanggol Kalikasan. (1998) Letter to DENR Secretary Victor Ramos, April 20.

Tarrow, Sidney. (1994) *Power in Movement*. Cambridge: Cambridge University Press.

Taylor, Peter J. (1994) "The State as Container: Territoriality in the Modern World System." *Progress in Human Geography* 18: 151–62.

Taylor, Peter J., Michael J. Watts, and R. J. Johnston. (2002) "Geography/Globalization." In *Geographies of Global Change: Remapping the World*, ed. R. J. Johnston, Peter J. Taylor, and Michael J. Watts, 1–17. Oxford: Blackwell.

Teehankee, J. C. (1993) "The State, Illegal Logging and Environmental NGOs in the Philippines." *Kasarinlan* 9: 19–34.

Thompson, Mark R. (1995) *The Anti-Marcos Struggle*. New Haven: Yale University Press.

Tolentino, Willy. (1994) Letter to DENR Secretary Angel Alcala, March 23.

Top, Gerhard van den. (2003) *The Social Dynamics of Deforestation in the Philippines*. Copenhagen: Nordic Institute of Asian Studies.

Touraine, Alain. (1981) *The Voice and the Eye*. Cambridge: Cambridge University Press.

Toye, John. (1993) *Dilemmas of Development: Reflections on the Counter-Revolution in Development Economics*. Second Edition. Oxford: Blackwell.

Tronto, Joan C. (1994) *Moral Boundaries: A Political Argument for an Ethic of Care*. New York: Routledge.

Turner, B. S. (1993) "Outline of a Theory of Human Rights." *Sociology* 27, no. 3: 489–512.

Tvedt, T. (1998) *Angels of Mercy or Development Diplomats? NGOs and Foreign Aid*. Oxford: James Currey.

Unger, Danny. (1998) *Building Social Capital in Thailand*. Cambridge: Cambridge University Press.

United States Agency for International Development (USAID). (1996) *1996 Annual Partners' Conference: Presentations, Volume 2*. Manila: USAID.

——. (1997) "Evolution of a PVO Co-Financing Program." Washington: USAID mimeo.

Utting, Peter, ed. (2000a) *Forest Policy and Politics in the Philippines*. Quezon City: Ateneo de Manila University Press.

——. (2000b) *Business Responsibility for Sustainable Development*. UNRISD Occasional Paper Number 2. Geneva: UNRISD.

Vandergeest, Peter, and Nancy L. Peluso. (1995) "Territorialization and State Power in Thailand." *Theory and Society* 24: 385–426.

Vaux, Tony. (2001) *The Selfish Altruist*. London: Earthscan.

Verian, R. (1998) "The Island of Coron and Its People." *Suhay* 2, no. 4 (October–December): 5–8.

Vitug, Marites Dañguilan. (1993) *The Politics of Logging: Power from the Forest*. Manila: Philippine Center for Investigative Journalism.

Walker, R. B. J. (1990) *Inside/Outside*. Cambridge: Cambridge University Press.

Walzer, M. (1994) *Thick and Thin: Moral Argument at Home and Abroad*. Notre Dame: University of Notre Dame Press.

Wapner, Paul. (1996) *Environmental Activism and World Civic Politics*. Albany: State University of New York Press.

Weekley, Kathleen. (1996) "From Vanguard to Rearguard." In *The Revolution Falters*, ed. Patricio N. Abinales, 28–59. Ithaca: Southeast Asia Program, Cornell University.

Weisbrod, Burton. (1977) *The Voluntary Nonprofit Sector: An Economic Analysis*. Lexington, Mass.: Lexington Books.

——. (1998) "The Nonprofit Mission and Its Financing." In *To Profit or Not to Profit*, ed. Burton Weisbrod, 1–22. Cambridge: Cambridge University Press.

Winichakul, Thongchai. (1994) *Siam Mapped*. Honolulu: University of Hawaii Press.

Wolfe, Alan. (1989) *Whose Keeper? Social Science and Moral Obligation*. Berkeley: University of California Press.

Wurfel, David. (1988) *Filipino Politics: Development and Decay.* Ithaca: Cornell University Press.

www.haribon.org.ph. Accessed on 11 July 2001.

Yearley, Steven. (1996) *Sociology, Environmentalism, Globalization.* London: Sage.

Yeung, Wai-chung. (1998) "Capital, State and Space: Contesting the Borderless World." *Transactions of the Institute of British Geographers* 23: 291–309.

Yin, Robert K. (1994) *Case Study Research: Design and Methods.* Second edition. London: Sage.

Zald, Mayer N., and John D. McCarthy. (1987) *Social Movements in an Organizational Society.* New Brunswick: Transaction.

Zelizer, Viviana A. (1997) *The Social Meaning of Money.* Princeton: Princeton University Press.

Zimmerer, Karl S., and Thomas J. Bassett, eds. (2003) *Political Ecology: An Integrative Approach to Geography and Environment-Development Studies.* New York: Guilford Press.

Interviews

Aguado, Marietta. (1997) Assistant Manager, Social Development Desk, San Miguel Corporation, Manila, 28 April.

Aguilar, Kudul. (1996) Chairperson, Tagbanua Foundation of Coron Island (TFCI), Coron Island, 19 October.

Albotra, Rodelia. (1997) Coordinator, Philippine Federation for Environmental Concern (PFEC), Manila, 23 April.

Alcala, Angel. (1997) Chairperson, Commission on Higher Education, former Secretary, Department of Natural Resources and Environment (DENR), and former Member, Board of Trustees, Haribon Foundation, Manila, 2 June.

Aldaba, Fernando. (1997) Executive Director, Ateneo Center for Social Policy and Public Affairs, Ateneo de Manila University, Manila, 27 May.

Alegre, Alan. (1997) Researcher, Ateneo Center for Social Policy and Public Affairs, Ateneo de Manila University, Manila, 2 May.

Ali M. K. (1997) Regional Manager, Plan International Philippines, Manila, 27 May.

Amos, Lourdes. (1997) Research and Documentation Officer, Philippine Association for Intercultural Development (PAFID), Manila, 18 April.

Aquino-Gonzales, Leonora. (1997) External Relations Officer, World Bank, Manila, 29 May.

Aquino-Elogada, Rachel. (1997) Assistant Program Officer, Asia Foundation, Manila, 8 May.

Araojo, Melvyn. (1996) Former Protected Area Supervisor, Department of Natural Resources and Environment (DENR), Naga City, 16 October.

Arquiza, Yasmin. (1997) Journalist and Editor, Bandillo ng Palawan [monthly Palawan newsmagazine], Manila, 25 April.

Austria, Isabelita. (1997) Supervising Forest Management Specialist, Community Based Forest Management Office (CBFMO), Department of Natural Resources and Environment (DENR), Manila, 9 June.

Austria, Joey. (1997) Board of Trustees, Philippine Association for Intercultural Development (PAFID) and Senior Officer, Indigenous Community Affairs Division, Department of Natural Resources and Environment (DENR), Manila, 29 May.

Bagayas, Rodolfo. (1997) Stakeholder Desk Officer, San Miguel Corporation, Cebu City, 23 May.

Balinhawang, Sammy. (1996) Area Coordinator (Luzon), Philippine Association for Intercultural Development (PAFID), Didipio (Nueva Vizcaya), 4 October.

Balungay, Mayrose. (1996) Administrative Officer, Haribon Foundation, Manila, 30 October.

Banzuela, Raul. (1997) Deputy Executive Director, Philippine Partnership for the Development of Human Resources in Rural Area (PhilDHRRA), Manila, 20 May.

Bayabos, Reynaldo. (1997) Assistant Director, Community Based Forest Management Office (CBFMO), Department of Natural Resources and Environment (DENR), Manila, 9 June.

Belen, Angelo Ruel. (1996) Liaison Officer and Community Organizer, Philippine Association for Intercultural Development (PAFID), Coron Island, Palawan, 20 October.

Braganza, Gilbert. (1996) Researcher, Environmental Research Division (ERD), Manila, 21 September.

Braza, Jun. (1997) Monitoring Appraisal Assistant, Canadian International Development Agency, Manila, 29 April.

Burillo, Rafaela. (1996) Community Organizer, Haribon Foundation, Manila, 29 October.

Calanog, Lope. (1997) National Co-Director, National Integrated Protected Areas Program, Manila, 6 June.

Casals, Camilo. (1997) Program Coordinator, Australian Agency for International Development, Manila, 9 May.

Chua, Sherah. (1997) Development Officer, British Embassy, Manila, 29 April.

Constantino-David, Karina. (1997) President, Caucus of Development NGO Networks (CODE-NGO) and Professor, Community Development, University of the Philippines, Manila, 7 May.

Copa, Sergio. (1996) Chairperson, Livelihood Committee, Samahan ng Mangingisda at Mamamayan ng Binabalian (SAMMABI), Bolinao, 26 November.

Dacanay, Lisa. (1996) Assistant Vice-President, Philippine Rural Reconstruction Movement (PRRM), Manila, 23 September.

De la Castro, Tony. (1997) Country Director, Conservation International, Manila, 8 May.

de los Reyes, Judith. (1996) Accountant, Haribon Foundation, Manila, 30 October.

Derige, Ramon. (1997) Coordinator, Upland NGO Assistance Committee (UNAC) and Director of Operations, Philippine Business for Social Progress (PBSP), Manila, 30 May.

De Vera, Dave. (1996 and 1997) Executive Officer, Philippine Association for Intercultural Development (PAFID), Manila and Coron Island, January 1996, 30 September 1996, 23 October 1996, and 18 April 1997.

Dizon, Jose Carlos Albert (1996) Community Organizing Development Program Coordinator, Haribon Foundation, Manila, 6 November.

Dugan, Patrick. (1996 and 1997) Policy and Program Support Consultant, United States Agency for International Development (USAID), Manila, January 1996 and 19 May 1997.

Dumalagan, Alfie. (1996) Chairperson, Samahang Mangangalaga ng Bukid (SMB), Naga City, 15 October.

Factoran, Fulgencio. (1997) President, Gaia South and former Secretary, Department of Natural Resources and Environment (DENR), Manila, 21 May.

Ferrer, Sam. (1997) Executive Director, Green Forum, Manila, 22 April.

Ganapin, Delfin. (1997) Undersecretary for Environment and Programs Development, Department of Natural Resources and Environment (DENR) and former Coordinator, Philippine Federation for Environmental Concern (PFEC), Manila, 2 June.

Garchitorena, Victoria. (1997) Executive Director, Ayala Foundation, Manila, 4 June.

Gasgonia, Donna. (1997) Execututve Director, Foundation for the Philippine Environment (FPE) and Member, Board of Trustees, Philippine Association for Intercultural Development (PAFID), Manila, 9 June.

Gomez, Edgardo (1997) Executive Director, Marine Science Institute (MSI), University of the Philippines, Manila, 6 June.

Granert, Aida. (1997) Development Officer, Soil and Water Conservation Foundation, Cebu City, 22 May.

Guiang, Ernie. (1996) Deputy Chief, National Resources Management Program, Department of Natural Resources and Environment (DENR) and Development Alternatives Inc. (DAI), Manila, January.

Guidote-Alvarez, Cecile. (1997) Communications Director, Earthsaver's Movement, Manila, 2 May.

Hawes, Gary. (1997) Program Officer, Ford Foundation, Manila, 3 June.

Holopainen, Jukka. (1997) Membership Officer, Haribon Foundation, Manila, 24 April.

Isberto, Ester. (1997) Executive Director, NGOs for Integrated Protected Areas (NIPA), Manila, 9 June.

Iscala, Nicanor. (1997) Supervising Forest Management Specialist, Community Based Forest Management Office (CBFMO), Department of Natural Resources and Environment (DENR), Manila, 9 June.

Kho, Demetrio. (1997) Chief Ecosystems Management Specialist, Region 7, Central Visayas, Department of Natural Resources and Environment (DENR), Cebu City, 22 May.

King, Belen. (1997) Member, Board of Trustees, Haribon Foundation, Manila, 19 May.

La Viña, Tony. (1997) Undersecretary for Legal and Legislative Affairs, Department of Natural Resources and Environment (DENR), Manila, 5 May.

Lim, Liza. (1996) Researcher, Institute of Social Order (ISO) and University of Hawaii, Manila, January.

Luna, Maria Paz "Ipat." (1996) Tanggol Kalikasan Program Coordinator, Haribon Foundation, Manila, 30 October.

Lynch, Damon. (1997) Sustainable Development Policy Specialist, Center for Alternative Development Initiatives (CADI), Manila, 6 June.

Magno, Lisa. (1997) Program Manager, United States Agency for International Development (USAID), Manila, 30 May.

Mangulatron, Ernest. (1996) Membership Officer, Haribon Foundation, Manila, 7 November.

Mendoza, Angel. (1996) Chairperson, Livelihood Committee, Samahang Mangangalaga ng Bukid (SMB), Naga City, 15 October.

Mercado, Orlando. (1997) Senator, Philippines Senate, Manila, 10 June.

Mesa, Fermin. (1996) Community Organizer, Haribon Foundation, Manila, 29 October.

Morales, Horacio. (1997) President, Philippine Rural Reconstruction Movement (PRRM), Manila, 30 May.

Navarro, Benjamin. (1996) Project Manager, Philippine Association for Intercultural Development (PAFID), Didipio (Nueva Vizcaya), 4 October.

Nozawa, Cristi. (1996 and 1997) Special Programs Coordinator and former Executive Director, Haribon Foundation, Manila, 28 October 1996 and 30 April 1997.

Osteria, Trinidad. (1997) Director, Social Development Research Center, De La Salle University, Manila, 9 May.

Ovara, Budge. (1996) Institute of Social Order (ISO), Manila, 8 January.

Perez, Domingo. (1996) Chairperson, Anduyog Federation and Captain, Barangay Lugsad, Naga City, 15 October.

Perez, Francis. (1997) Deputy Director, Tambuyog, Manila, 5 May.

Plantilla, Annabelle. (1997) Member, Board of Trustees, Haribon Foundation, Manila, 31 May.

Racelis, Mary. (1997) Assistant Representative, Ford Foundation, Manila, 25 April.

Reoma, Gil. (1997) Coordinator, Green Forum, Manila, 22 April.

Resurrecion, Noel. (1996) Project Officer and Community Organizer, Haribon Foundation, Naga City, 14 October.

Reyes, Ed. (1997) Secretary General, Communicators League for Environmental Action and Restoration (CLEAR) and President, Saniblakas ng Taongbayan, Manila, 3 June.

Rice, Delbert. (1996) Chairperson of the Board of Trustees and former Executive Officer, Philippine Association for Intercultural Development (PAFID), Manila, 9 October.

Rivero, Cris. (1996) Protected Area Supervisor, Department of Natural Resources and Environment (DENR), Naga City, 16 October.

Roa, Jeruel. (1997) Manager, Corporate Affairs Office, San Miguel Corporation, Cebu City, 23 May.

Rodriguez, Raul. (1997) Manager, Social Development Desk, San Miguel Corporation, Manila, 28 April.

Romero, Gene. (1996) Researcher, Environmental Research Division and former Policy Analyst, Haribon Foundation, Manila, 3 December.

Romero, Leandro. ["Andrei."] (1997) Land Tenure Program Officer, Philippine Association for Intercultural Development (PAFID), Manila, 18 April.

Roque, Celso. (1997) President, Kabang Kalikasan ng Pilipinas, Director of Policy Asia/Pacific Program of World Wildlife Fund (WWF), former Undersecretary Department of Natural Resources and Environment (DENR), and former Executive Director, Haribon Foundation, Manila, 6 June.

Rosales, Godo. (1996) Former Vice-Chairperson, Samahang Mangangalaga ng Bukid (SMB), Naga City, 15 October.

Roxas, Elizabeth. (1997) Executive Director, Environmental Broadcast Circle, Manila, 3 June.

Rubio, Enrico. (1997) Manager, Center for Clean Technology and Environmental Management and Philippine Business for the Environment (PBE), Manila, 28 April.

Saniel, Joan. (1997) Lawyer, Environmental Legal Assistance Center, Cebu City, 23 May.

Serrano, Isagani. (1996) Deputy President, Philippine Rural Reconstruction Movement (PRRM), Manila, 23 September.

Severino, Howie. (1997) Environmental Desk Coordinator, Philippine Center for Investigative Journalism, Manila, 25 April.

Supetran, Amelia. (1997) Project Leader, Environmental Management Bureau, Department of Natural Resources and Environment (DENR), Manila, 9 June.

Tan, Julio. (1997) Director for Institutional Development, Foundation for the Philippine Environment (FPE), Manila, 4 June.

Teunissen, Hans. (1997) Counselor, Royal Netherlands Embassy, Manila, 27 May.

Tindungan, Margareth. (1996) Administrator, Philippine Association for Intercultural Development (PAFID), Manila, 8 November.

Tolentino, Aurora. (1997) Executive Director, Philippine Business for Social Progress (PBSP), Manila, 30 May.

Tolentino, William. (1996) Land Tenure Coordinator and former Executive Officer, Philippine Association for Intercultural Development (PAFID), Manila, 8 October.

Tongson, Ed. (1997) Executive Director, Haribon Foundation, Manila, 8 October 1996 and 19 April.

Turion, Rene. (1996) Community Organizer, Haribon Foundation, Manila, 29 October.

Vale, Ronnie. (1996) Chairperson, Samahang Pangkalikasan ng Cawaynan (SPC), Naga City, 14 October.

Van Engelen, Herman. (1997) Director, Water Resources Center, University of San Carlos, Cebu City, 23 May.

Vargas, Eric. (1996) Area Coordinator (Mindoro), Philippine Association for Intercultural Development (PAFID), Manila, 8 October.

Villavicencio, Veronica. (1996) Director of Grants Program, Foundation for the Philippine Environment (FPE), Manila, January.

Vitug, Marites Dañguilan. (1997) Writer and Journalist, Philippine Center for Investigative Journalism, Manila, 22 April.

Walpole, Peter. (1996 and 1997) Executive Officer, Environmental Research Division, Manila, 21 January 1996 and 5 June 1997.

Walsh, Thomas. (1997) Senior Programs Officer, Asian Development Bank, Manila, 5 June.

Wijangco, J. Ernesto. (1996 and 1997) Program Manager, United States Agency for International Development (USAID), Manila, January 1996 and 2 June 1997.

Wilkinson, Gordon. (1997) Senior Social Development Specialist and NGO Coordinator, Asian Development Bank, Manila, 5 June.

Zuniga, Maria Luisa. (1997) Program Coordinator, Sagip Pasig Movement, Manila, 29 May.

Index